ERGONOMIC
OF
OLD AGE HOMES AND HEALTH

About Authors

Mrs. Shewanti N. Kashyap did her B.Sc. (H.Sc) and M.Sc. (Home Management from Nagpur. She joined the Department of Family Resource Management at Govind Ballabh Pant University of Agriculture and Technology, Pantnagar for the degree of Doctor of Philosophy as a sponsored candidate from Rajendra Agricultural University Pusa (Bihar), Samastipur.

Dr, Promila Sharma is Professor in the discipline of Family Resource Management, G.B.P.U. of Ag. & Tech. Pantnagar. She is having vast experience of research in Erogonomics field.

Dr. A.K. Shukla is Professor, former Dean, College of Basic Sciences. He is working as Professor and Head of Department of Maths, Statistics and Computer Application Science for last 15 years. Dr. Shukla had guided several Ph.D. and M.Sc. students of Department of Resource Management on topics like human factor and ergonomics, women work and stress, physiological and psychological cost of work of hill women, injury hazards ill health and fatigue among rural and urban women workers etc. Dr. Shukla had deep knowledge on giving shape to relational statistics on social researches. He has commendable knowledge on giving shape to researches like action research/experimental/descriptive/exploratory research.

ERGONOMIC OF OLD AGE HOMES AND HEALTH

Shewanti N. Kashyap
Promila Sharma
and
A.K. Shukla

CONCEPT PUBLISHING COMPANY PVT. LTD

NEW DELHI-110059

ISBN-13 : 978-81-8069-997-9

First Published 2014

Published and Printed by

Concept Since 1974

Concept Publishing Company Pvt. Ltd.
A/15-16, Commercial Block, Mohan Garden
New Delhi-110059 (India)
T: 25351460, 25351794, *F*: 091-11-25357109
E: publishing@conceptpub.com, W: www.conceptpub.com
Editorial Office: H-13, Bali Nagar, New Delhi-110 015, India.

Cataloging in Publication Data–Courtesy: D.K. Agencies (P) Ltd. <docinfo@dkagencies.com>

Kashyap, Shewanti N.
Ergonomic of old age homes and health / Shewanti N. Kashyap, Promila Sharma and A.K. Shukla.
p. cm.
Includes bibliographical references (p.) and index.
ISBN 9788180699979

1. Old age homes—India—Kumaon Haldwani Block. 2. Old age homes—India Garhwal. 3. Older people—Institutional care—India—Kumaon Haldwani Block. 4. Older people—Institutional care—India—Garhwal. 5. Human engineering India—Kumaon Haldwani Block. 6. Human engineering—India—Garhwal. I. Sharma, Promila, 1955-, joint author. II. Shukla, A. K. (Professor of mathematics), joint author. III. Title.

DDC 362.61095451 23

Dedicated

to

Elderly People

Preface

The term "ergonomics" is derived from two Greek words: "ergo," meaning work, and "nomoi," meaning natural laws. Ergonomists study human capabilities in relationship to work demands. As the body ages it needs special ergonomic considerations. The demand for old age homes are increasing, knowledge on furniture, furnishing, locations, storage facilities, infrastructure and other fittings and fixtures essential. These topics are discussed in detail. Since the harmful effects of housing are being realized the importance of Ergonomics are given due consideration. This book discusses those needs and how to adjust your environment to aid your ageing body in a very comprehensive way and expected to serve as a good reference book and useful guide. By learning more about the discipline of ergonomics, individuals can begin to understand how to improve the surroundings of older adults and, in turn, help maintain their quality of life. By taking into account the needs of the elderly population, and also the extent of their cognitive and physical abilities and disabilities, professionals and health care workers can all help in designing and organizing their environment in manners that help improve and maintain their quality of life of old age people.

Shewanti N. Kashyap
Promila Sharma
and
A.K. Shukla

Acknowledgements

We never knew this colossal task would seem so simple and attainable by the expert, timely and invaluable sparks of guidance and wisdom of Advisory Committee. We extend our heartfelt thanks to Dr. S.K. Rastogi, Professor and Head, Department of Veterinary Physiology, Dr. S.P. Singh, Professor and Head, Department of Agriculture Economics, and Dr. S.S. Gupta, Professor, Department of Civil Engineering for their constant encouragement, constructive suggestions and critical evaluation of the manuscript.

We are highly grateful to Dean, College of Home Science and Dean, Post Graduate Studies for providing the necessary facilities for conducting the research. We are immensely grateful to Mrs. Rajni Lakhani, Scientist, OBM Division, CBRI, Roorkee, for providing the most relevant document—"Skid Resistance Polycemtiles", and Mr. V.K. Verma, Asstt. Professor, College of Technology, Department of Civil Engineering for his valuable help in testing the "Flooring Material".

Last but not the least, we are highly thankful and obliged to Dr. P.C. Gope, Professor and Mr. V.K. Singh, Associate Professor, Department of Mechanical Engineering for "Designing Chair" which will be a great contribution for the elderly people.

Shewanti N. Kashyap
Promila Sharma
and
A.K. Shukla

Contents

List of Tables

List of Figures

1

Introduction

The year 1999 was selected as the International Year of Older Persons by United Nations, partly due to the fact that the ageing population is a universal fact touching both developed and developing countries. Furthermore, with all the research and development in the health sector, man is living longer. Living longer may sound as a good thing, but if one has to add extra years to his life span, with years of dependency and incapacity, it will not be good anymore. This will then suggest that most of the people who are in their 1990s are invalids, and this has been called "the failure of our success" *(Coni, et al., 1992)*. The elderly people are changing in many respects and these can be categorized as morphological and functional. Today, they are healthier than the previous generations, with all the opportunities made available to them through developing technology, with better education and better financial support. They are the people who demand more and who still often have an active role to play in society.

Demographic ageing which was until recently perceived to be a concern of the industrialized countries only, has become a global phenomenon. With better medical care and improved nutrition, people are living longer. In India too, life expectancy has improved resulting in increasing population of older persons. This population shot up from 12 million in 1901, to 19 million in 1951 and 77 million in 2001. It is anticipated that by 2021 it will reach 137 million. India now has the second largest aged population in the world.

Indian society provides a congenial set of conditions for a physically comfortable and emotionaliy satisfying old age. It is a social expectation in India that the adult sons will take care of their aged parents and the daughters take care of the parents in case the parents do not have sons *(Vatuk, 1990)*. But all this is changing. There has been a marked change in the treatment of the elderly compared to earlier periods. Many of the recent studies emphasized the transition in the role and status of the elderly Indians from pre-industrial society to the existing industrial social order. In this kind of social situation, conflict and problems of adjustment has become a normal feature of the Indian families. Many of the elderly parents are compelled to leave their children and stay in old age homes. However, in more recent times, institutions such as old age homes are sharing the traditional role of the family. The old age homes which were a rarity few decades back have recently spread across the country, a fact that indicates the growing rift between the generations. It is sometimes asserted that the absence of any such care will cause adjustment problems, loneliness and depression in the milieu of old age homes as the elderly persons are habituated to staying with their family members. The residents of the old age homes would find it difficult to adapt to the setting of the old age homes, which lacks the warmth and security of the family environment. *Mishra (1993)* mentioned in her study that the children of the elderly residents did not object to and expressed no sadness at the decision of their parents to reside in old age home rather they felt relieved.

Housing must fulfils the needs for the daily activities of the elderly, and more precisely it must give them: satisfaction, security, comfort, and independence or at least one of them. The house interior depicts three main basic functions as usability, safety and attractiveness. Comfort, daily life activities, ease of mobility, and independence are listed under the group of usability. They are all affected by the accessibility of the environment. Shelter, security, safety, and privacy are listed under the group of safety. Aesthetics, pleasure in use, and quality belong to the group of attractiveness. Comfort includes the feeling of thermal comfort (temperature, humidity, and drafts), adequate lighting level and colour vision comfort, hearing comfort, and physical comfort. Daily life activities include: cooking and eating; sleeping and resting; grooming; dressing and undressing; use of shower, bath, toilet, and wash basin; operating doors and windows; washing clothes; washing dishes; carrying objects; moving furniture; operating thermostat; reaching for high objects; chair comfort (sitting, standing up); counter convenience (storage); stair and ramp use (elevator operation); finger and manual tasks (sewing, writing, playing cards); getting in and out of bed; hearing and viewing things; using household appliances (vacuum, iron); performing hobbies (plants, knitting, needlework...); watching TV, listening to radio, recorders; wheel chair mobility; operating a telephone (accessibility, convenience); etc. *(Demirbilek, 1996b).* Ease in mobility includes all the movements (with their degree of easiness) that a resident can make while circulating in the house. Privacy, security and safety are interrelated. They deal with available systems and designs preventing and helping from any accident that may occur. A well-designed house will ensure greater use, safety, security, and privacy.

Sheehan (1992) stated that the physical design of housing interiors plays a major role in influencing the quality of life of all elderly residents. Moreover, poor functional design is said to cause a lack of confidence in users and create a sense of frustration. Poor ergonomic design can cause pain, discomfort and injury; poor aesthetic image can create stigma and decrease the pleasure in using spaces or objects *(Allan et al., 1996).* Falls are common among the elderly and often lead to loss of independence or even death. A home and personal safety evaluation may uncover many high-risk conditions. The team may learn about near misses and remove hazards. Some risks are easily remedied, but others may require major home modifications. Nevertheless, there are indications that physical conditions do influence the physical development, social attributes and family attributes. Physical aspects frequently determined the way in which a house will be constructed and later furnished. The quality of housing can enhance or diminish the well being of individual and families as that of the entire community. Hence, this is a major field where such improvement could be achieved by considering the human factors involved in it. There home ergonomics is becoming very important amongst home scientists, ergonomists, industrialists, builders and interior designers

(Varghese et al., 1989). Safety problems belong to the basic objectives of shaping the artificial environment of human life. Attention is focused mostly on road accidents and work environment, less attention is paid to dangers arising in housing.

Ergonomics is defined as the study of the anatomical, psychological and physiological aspects of man in his working environment, with the objective of optimizing human safety, health, comfort and efficiency. Traditional ergonomics, which has grown from labour sciences, dealt mainly with adjusting machines tools and devices, as well as the material environment of work to the needs and possibilities of people enjoying full mental and physical ability. Modern ergonomics increasingly deals with problems of adjusting the environmental conditions to the needs of people of limited ability as well.

Rationale of the Study

Some factors like outcome of increased urbanization have adversely affected older people. One is the migration of the younger adults of the family to towns and in some cases to foreign countries, has abandoned their parents back home who may be reluctant to migrate, because of difficulty in adjusting to the lifestyle in new set-up. Besides it, increased participation of women in paid employment outside the home is another factor attributing to increase old age homes in India. Help Age India has sponsored the construction and maintenance of old age homes in India. These homes cater to the needs of those elderly who are unable to live by themselves and for those who have been abandoned by the family or are neglected by their children. These old age homes provide and cater to the various needs of the elderly so that they can spend the "evenings of their lives" with dignity and respect and not feel a burden to the society. There are over 800 old age homes all over India.

The environment must be designed to ensure no harm, injury or undue risk. Loss of self-confidence as a result of infirmity means that window bars, doors and locks must be designed to offer security without becoming obstacles. Provision of a functional environment for the physically limited demands easily accessible doors and rooms to manoeuvre when bending, sitting and standing. Accidents at (Home and Professional) work are more frequently to blame on human fault, incompetence or ignorance than on the material environment. Complete elimination of accidents is almost impossible, but their number and effects may be minimized by ergonomic education and formation of ergonomic consciousness in society.

The basic components essential in providing an appropriate residential environment for this particular population, whose distinguishing characteristics include impairment of basic functioning ability, lack of self-confidence, feeling of isolation and loneliness and lost identity. The design and dimension of the house affects the physiological and psychological comfort and discomfort thus may be estimated and need is to find out the ways to reduce it. The ill design of the house may produce degenerative changes in body. Many studies were conducted on housing and elderly people but not much study are undertaken on ergonomic housing and elderly people. There is an urgent need to conduct research on important issues that is the requirement of elderly people for functional, comfortable, safe housing and environment. Keeping the above issue, in mind, it is proposed that the unit for elderly should be ergonomically comfortable and should be constructed on certain guidelines. An ergonomically designed unit is friendly to human body and the dimensions of house are harmoniously matched with each other in ergonomically designed houses, hence, the present study entitled "Ergonomic assessment of old age homes and related health hazards on elderly people" was planned with the following objectives to be fulfilled.

Objectives

1. To find out the general profile of elderly people.
2. To study the existing construction and dimensions of old age home in the light of ergonomic principles.
3. To study the infrastructure facilities available in old age home.
4. To find out the satisfaction derived by the elderly people

from the old age home.

5. To study related health hazards faced by the elderly people in old age home.
6. To develop a guidelines for functional constructive unit for elderly people.

Hypothesis

1. The age and sex of elderly people have no relationship with occurrence of accidents.
2. Anthropometrical measurements of elderly people have no relationship with dimensions of (a) storage areas and (b) furniture.
3. Anthropometrical measurements of elderly males have no relationship with anthropometrical measurements of elderly females.
4. Dimensions of old age homes have no relationship with National Building Codes.
5. Dimensions of furniture have no relationship with Indian Standard Codes.

Limitations

1. The study was limited to the sixty elderly people of four old age homes of two regions of Uttarakhand State.
2. Since, the housing and ergonomics is a vast area of research only selected feature of old age homes, *i.e.* storage spaces, environmental parameters, fittings and fixtures, and furniture were taken into consideration.

2

Review of Literature

A systematic and thorough review of studies related to the problem under study, help in analyzing the problem, its historical status and its development and current status. The available review of literature was divided in various heads:

2.1. Studies related to construction and dimensions of old age homes.
2.2. Infrastructure facilities available in old age homes.
2.3. Studies related housing satisfaction.
2.4. Studies related to health hazard/health problem.
2.5. Guidelines for functional constructive unit for old age homes.

2.1 Studies related to construction and dimensions of old age homes

Howe (1992) stated that the best way to ensure that the concerns of the elderly are met, and that local plans reflect their needs and perspectives, was to have them actively participate in the local planning process, with regular meetings.

Woudhuysen (1993) claimed that elderly people, besides responding to questionnaires and attending to focus groups, should also work in teams with designers, entering early and directly into the design process.

Allan et al. (1996) reported that the ideas and comments of old people certainly play an important role in the building design process, and stated that there is a need for much more involvement of elderly users at all levels in the design process.

Brink (1996) pointed out that most dwellings are not "senior-friendly" or barrier-free, and that those dwellings are designed without considering even the basic requirements of elderly residents, resulting in their exclusion from everyday life.

Cavanagh (1996) pointed out that the opinion of the elderly themselves related to the design itself is never or rarely considered. The study carried out by the Women's Design Service in which old women were interviewed. In these interviews, the different needs of people; their preferences in using interior house spaces; and their housing requirements, were inquired. The primary suggestions of the elderly women in this study were as follows:

- Social housing should be built with space standards suitable for wheelchair users;
- Heights and standard fixtures and fittings (taps, cupboards...) should be usable by everybody including people having restriction in mobility and dexterity;
- Particular attention should be given to designing features in houses that would help in reducing the risk of accidents (for example: falls); and
- More clear information and cues should be available to know what to do in case of a fire in the house.

Kose (1996) stated that the dwellings designed with the universal design philosophy are called differently in various countries: 'Life Span Dwelling' in Norway, 'Universal Design, Adaptable Housing, and Flexible Housing' in the UK, 'Life Time Home, Easy Adaptable' in Sweden, and 'Elderly Dwelling' in Denmark. 'Smart Homes' are new trials for elderly homes where technology is used widely for maximum comfort and safety.

Morini and Pomposini (1996) recommended the recognition of the value of user participation in design and the endurances that design guidelines and advise are based on user data (from representative samples of all potential users). These authors believed that looking at alternatives for adaptability that allow residents to age in place during the design process was a less expensive solution. Furthermore, they proposed that the housing design process should be enriched with real interventions (of end users), in which the evaluation of every case with specific constraints of tradition, habits, funds, time and country laws would be made.

Richardson and Poulson (1996) believed that the application of methods, tools, techniques, and knowledge while dealing with questions of usability, would improve product and service quality. They also believed that it would also contribute to increased social integration, the quality of life, and independent living for ameliorating social consequences, and improving product and service market expectations. According to them, designing for the elderly and disabled people was designing for everyone.

Coleman R. (1997) pointed out the importance of consulting elderly people during the design process, because they claimed that elderly people are surrounded by things which do not work well for them, or that they simply cannot find the things that they want. As ageing affects everybody everyday, designers were in fact designing for their own future. He added that "to do that successfully, a new collaboration between old and young" was needed. *EUFRP (1997)* is one of them, where years of experience, knowledge and wisdom of the elderly were used to create new designs in the field of virtual communication and web use. *Wolff et al. (1997)* conducted researches oriented towards the construction of housing for the elderly identified: senior housing, sheltered housing, nursing houses, community dwellings, and so on for old age people.

Boess et al. (1999) conducted a study on bathrooms where elderly participants were preparing mood-boards in sessions, using a ready-made toolkit, while being encouraged to add their own drawings and comments to their boards. This was done to let people act as 'openly' as possible in the research; i.e. not constraining them with questions that are too tight (closed), and at the same time not leaving them on the other extremity with questions that are too open. These mood-boards were discussed during the sessions to let each participant describe and justify his/her own design decisions.

Mishra (1993) Bagga (1997) Dandekar (1996) Nalini (1997) Chadha and Kanwar (1998) Ranjan et al. (1999) have looked at the structure of old age homes, life of the inmates, and their level of satisfaction or dissatisfaction, loneliness, depression and family linkage of the inmates.

Scott Woodcock (2005) pointed out that the most accidents occur in the home, but good building design can help to achieve a much safe and healthier living environment. The location of a house, construction methods, materials, finishes, appliances and maintenance all influence home safety. Incremental improvements to occupant health and safety can be achieved at every stage of the building design and construction process.

2.2 Studies related to infrastructure in old age homes

Kose (1993) accused architects and designers for not being prepared, as yet, to accept design requirements for elderly people, which he listed as follows: elimination of level differences (including the elimination of door sills); installation of handrails, installation of door level handles, installation of larger switches, installation of easily operable facilities, etc. There are certainly many other requirements that can be added to this list, such as the provision of adequate lighting avoiding large contrasts :

1. Ambient illumination that changes the light level gradually,
2. Provision of sliding dimmers (because visual impairment is progressive),
3. Pin-up wall fixture lamps for reading (suspended lights are not efficient here),
4. Reduced contrasts between lamp shades and their background (by using translucent shades = glare free light sources),

5. Additional lighting to warm and guide the elderly,
6. Use of uni-directional lighting to indicate the approach and the height of steps,
7. Use of reflective matt finishes, high reflectance laminates and tiles inside the cabinets (cupboards),
8. For the aged kitchen needs a lot of light everywhere, provided that the wall finishes and the cabinetry remain reflecting.

Salmon (1993) pointed out that there are certainly many other requirements that can be added to this list such as the provision of adequate lighting avoiding large contrasts illuminated and well located light switches, visual as well as auditory alarm systems, differentiation of wall and floor surface textures, colour coding, induction loops to assist hearing aids, sound insulation, rounded corners and edges, sliding doors (particularly for cabinet and cupboard doors), contrasting stripe on the edges of the treads of any stairs, handrails extended beyond the top and the bottom of stairs etc.

Three areas of concern of the visual loss in old age *(Helms, 1980; Kellog et al., 1986; Raschko, 1991;* and *Sturges, et al. 1994)* findings were as under :

Effects on the eyes	*Effects on activities and design*
Presbyopia	
This is the gradual reduction in the eye's ability to focus on close objects (accommodation). It begins around the age of 40 *(Helms, 1980)*. Eye glasses can be used to improve vision, but more light (for reading or doing detailed tasks) is needed.	Luminous keyholes, pay phones, switches and wall plates may be helpful *(Kellog et al. 1986)*. It can have significant negative influences on performance of some aspects of personal activities of daily living such as fastening buttons, causing functional impairments *(Sturges et al. 1994)*. Movements involving focusing, distances, steps, door sills, and pathway obstructions become hazardous *(Kellog et al. 1986)*. Additional lighting can be designed to warn and guide the elderly.

(Contd...)

Effects on the eyes	*Effects on activities and design*
Filter effects	
The yellowing cornea of the elderly acts as a colour filter and prevents the blue colour from travelling after the retina and from reaching the brain. The ability to see blue and mixtures containing blue is reduced. This decline in contrast sensitivity deprives the older person of the simultaneous contrast so important to colour-vibrancy *(Kellog et al., 1986).*	An aged eye cannot see colours well. This makes elderly people more sensitive to glare. Cataracts (due to opacities in the crystalline lens of the eye) produce an opaque condition for the eye lens which obstruct vision. The designer must then be aware that older people need primary hues of high chrome in contexts of high contrast and should avoid using blues (for example, it must be avoided to have signs and signage's blue on green background or the contrary).
Changing visual fields	
The reduction in the visual field makes it difficult to see out of the corner of the eye. Both the directions of view and the visual field change. The perception of depth is reduced both in the sense of distance and three-dimensionality.	The sharpness of edges is blurred, cues to depth are weakened. As the peripheral vision declines, the aged is easily disoriented, unaware of the objects that are not in direct view.

Bailey (1995) pointed out that designers should ensure that features such as door frames, door handles, steps, stairs and walkways should be well-distinguishable by visual contrast. These visual contrasts can be achieved either through brightness or colour differences to make the key features more perceptible. Glare from luminaries, windows and shiny surfaces should be controlled in the house. Local adjustable task lighting can be used to meet the special needs of elderly people.

2.3 Studies related to housing satisfaction

Housing satisfaction is an area of research that has received considerable attention. Much of this literature is concerned with examining the types of factors that may have an influence on satisfaction. For example, among the factors found to affect housing satisfaction most strongly is inadequate space *(Kinsey and Lane, 1983).*

Bagga (1997) mentioned that the under-engaged and idle residents of old age homes feel more depressed and listless. *McCray and Day (1977)* reported that the differences in housing satisfaction have been found between urban and rural residents. The former are more satisfied with inside storage and inside appearance, and that latter are more satisfied with the structural quality of their units and where their units are located in relation to church and social activities. Health status may predispose individuals to be satisfied with most things, including their housing. Therefore, longitudinal studies would be very helpful in exploring the potential role of causation in the relationship. The findings of a study of three priority housing action areas in Liverpool suggest an association between capitalized housing improvements and higher levels of housing satisfaction. However, the authors caution that, although improving housing conditions may improve physical and mental health, housing improvement policies should be firmly entrenched in social policies that address the interacting elements that exacerbate social inequalities in health.

Chadha (1998) compared the level of loneliness between the institutionalized and non-institutionalized elderly using UCLA loneliness scale, they do not provide details of causes of loneliness, their subjective perception of feeling lonely and what strategies they employ to overcome it.

Ranjan et al. (1999) study indicates that the majority of the residents are "overwhelmingly" happy, but it has not investigated what makes them feel so satisfied in the old age homes.

2.4 Study related to health hazard/health problem

- According to *Pitt (1982)*, ageing is a progressive decline in function and performance which accompanies advancing years. Nearly in all species the female have a tendency to live longer than the male. As an average, women live 5 years longer than men and they tend to be a little younger than the men they have married. As a result, there are more widows than widowers *(Pitt, 1982)* and this is other important factor to be considered while designing old age homes.

Esmen (1985) stated that the air exchange in residential buildings is chiefly by infiltration, which is affected by the kind of construction, the age of the house, and the weather protection provided. Air exchange rates vary by season and are affected by weather conditions and lifestyle. Unless air-cleaning devices are present, the sole major mechanism for removal of non-reactive gases is air exchange.

Samet et al. (1988) stated that the risks of condensation, smells, other pollutants, and a lack of fresh air result from too little ventilation, although too much can also increase the risk of condensation. Overall reduction of indoor air pollution may be achieved through increased ventilation throughout the structure or in particular areas.

Lowry (1989) reported that the concentration of pollutants depends upon rates of production and removal of the source, and their dilution by ventilation. The exchange between indoor and outdoor air profoundly affects contaminant levels in a dwelling.

Walker (1990) published "Housing and Health" focuses on the difficulties of doing research on the topic and quoted that "we tread on difficult epidemiological terrain when we attempt to develop a statistically precise analysis of the inter-relationship between housing and health." Many of the most important dimensions, he noted, are not susceptible to measurement. Studies that attempt the difficult task of measurement must leave out many variables for practical reasons, must often use indirect indicators or measures, and must often ignore important variables which simply cannot be measured. He summarized the problem in the following way: Typically, epidemiologists have all kinds of statistical data on causative agents of disease. But epidemiology require more than knowledge of specific organisms or other primary detergents of disease and dysfunction. It equally requires knowledge of the contributing factors or secondary determinants – the community, the psychology of its people, their social and economic position – and consideration of the possibility that they may operate independently and singly or synergistically. Thus, any exploration of housing and health issues must include a reflection of socio-economic position – an amalgam of income, education and occupation – which can be rightly called a true generic "housing-health risk factor" *(Walker, 1990)*.

Collins (1993) reported that the healthy dwellings should have both kinds of ventilation. Adequate ventilation is imperative to discourage humidity from reaching unacceptable levels. Low air movement and poor building ventilation are also risk factors for heat illness. *Collins (1993)* stated that the housing must provide protection from extremes of temperature well as other environmental hazards. The thermal equilibrium of the body is affected by four environmental elements in the indoor climate: ambient air temperature, radiant temperature, air humidity, and air movement. The interaction of these factors generates thermal stress and can cause undue heating or cooling of the body.

Enquselassie et al. (1993) reported that the occurrence of noticeable shifts in indoor conditions are a health threat, particularly for vulnerable groups like the elderly, infants, the sick and the disabled and those who spend most of their time indoors. For example, temperature stress resulting from both daily and seasonal temperature changes has been shown to be related to coronary episodes in Australia. Consequently, researchers there advise that improvements in residential temperature control would avert temperature extremes that produce this kind of stress.

Mood (1993) concludes that in terms of air quality, ventilation can take on two different meanings: (1) the supply of fresh outdoor air to occupied regions of a building and (2) air circulation within a dwelling, not necessarily including the supplementation of any fresh air. *Mant (1993)* investigated that the respiratory and circulatory diseases are the most often associated with wintertime deaths. Rates of mortality for these diseases rise as the temperature drops. *Markus (1993)* stated that it is necessary to have a degree of air exchange to control the build-up of moisture and to provide fresh air. *Nicol (1993)* stated that a restricted range of temperature variation has been established as the source of both thermal comfort and satisfaction. Furthermore, comfort and satisfaction have been found to depend on the capacity to control temperatures and adjust to changes.

Barbaccia (1995) stated that the musculoskeletal dimensions, mechanical performance, flexibility of joints, muscle strength, gait speed, bone densities are all important factors in the physiological system and changes occur in these with ageing. Arthritis, heart disease, diabetes, difficulty with vision and hearing are more common in older people. These problems with muscles and joints as hip fractures really contribute to a decrease in activities of daily living and instrumental activities of daily living.

Hopton and Hunt (1996) found that an aspect of poor housing with dampness that significantly and independently associated with greater emotional distress. *Rudge (1996)* reported that there is an apparent association between high levels of mortality in winter and low temperatures. Some medical research suggests that mortality among old people peaks in winter are not the result of extremely low temperatures, but of a seasonal fall in mean temperature, regardless of the average annual temperature. Thus, relative temperature changes as opposed to absolute temperature levels appear to be involved in the health impact.

Gitlin (1998) reported that the design of the structural environment should support, rather than restrict a person's daily life. It also allows for the individual to meet physical, psychological and social needs, and promote a sense of well-being and safety.

Gill and colleagues (2000) concluded that there are more important features of home and personal safety inspections than simply checking for hand railings and throw rugs to prevent trips and falls. These include increasing muscle strength, improving gait and balance, teaching proper use of assertive devices, wearing correct footwear, and monitoring medications.

Ranzijn (2002) concluded that the extent of the person-environment fit determines the quality of life of the older people.

2.5 Guidelines for functional constructive unit for old age homes

Classification of disabilities: design needs *(Raschko, 1991)* showed the following details: Central field defect macular generation severe loss of acuity, colour discrimination greatly affected.

Sensory disability	*Characteristics*	*Needs*
• VISION		
1. Central field loss	Central field defect macular generation severe loss of acuity, colour discrimination greatly affected	Colour cueing may not help. Value cueing may help. Use of light and dark tones, Oversize lettering on signs, auditory and tactile cues.
2. Peripheral field loss	Usually developmental: glaucoma retinal diseases etc. Results in tunnel vision, causes poor orientation, confusion. Colour vision is not affected.	Colour cueing helpful. All other aids applicable.
3. Cataract corneal disease	Clouding, distortion or opacities of lens or cornea of the eye. Slight to severe effect on vision.	Strong colour cueing, auditory and tactile cues, Oversize letter on signs, diffuse lighting.
• Blindness	Total blindness or persons may only perceive light. If person is familiar with his environment may be very self-sufficient.	Raised letters on signs. Generally more space is needed in living unit & kitchen.
• Hearing Loss	May be mild to severe congenital, progress (proboscises due to disease) old age. May be neural (high tones) conductive (low tones) or flats (all tones), Ringing noise may be present may also affect balance.	Visual and tactile cues necessary, Signal light for emergency along with bed vibrator when sleeping.
• Tectile Loss • Touch • Temperature	Sensory deprivation often accomplished by motor loss, brought about by variety of disorders, usually accomplished by varying degrees of paralysis.	Will need protection from injury, insulation of exposed hot water pipes, water controls in kitchen and bathroom, need to avoid sharp and abrasive surfaces.
• Motor Loss	May be due to vestibular impairment usually accomplish paralysis spasms amputation. Person may use walking aids; move more slowly with reaction time.	Special care for furniture and furnishing selection for safety poor balance is not a deciding as to self-care.

(Contd.)

(Contd.)

• Incoordination	Can be caused by cerebral palsy, multiple sclerosis chorea, inhibit the performance of simple task can affect all ages.	May need assistive services, some condition trainable, special aids for disability, prostheses.
• Stamina	Cardiopulmonary disorders, servere hypertension, advanced age, activities that are painful due to varieties of disorders, e.g. herniate vertebral disc.	Depending upon degree of disabilities elderly may need assistance at home. Avoid stairs; break tasks down in duration, severity of effort. Avoid prolong aggravation of injury.
• Reaching and Lifting with ARMS	Persons with limited strength range of motion of arms many cause: bursitis, arthritis, paresis.	In moderate cases may be able to function in self-help environment. In sever cases, may need assistance.
• Handling • Fingering	Inability to perform hand movements, arthritis, congenital deformities, spasms contractures etc.	Special aids furnishing equipments to offset disabilities, grasp, twist, push-pull oversize handles, grasping surfaces reduce two hand operation and *vice versa*.
• Upper Extremities • Skill	Complete paralysis, absence of upper extremities often quadriplegia.	Require assistance, ability to self-help dependent on retention of other skills or to be trained. Puff system.
• Bending • Kneeling	Stiff joints, arthritis, casts, braces often causes difficulty in sitting and kneeling.	Usually able to take care of self. Storage of often used items in middle and upper ranges.
• Reliance on Wakling AIDS	Braces, prostheses, canes, walkers, many disabilities can cause arthritis, polio, and trauma.	Need special help for ambulation, special design features needed.
• Inability of use of Lower Extremities	Confined to wheelchairs, amputation etc.	Independent living possible must have supportive devices, needs adaptation accessible environment.
• Extreme in Size	Dwarfism, gigantism, small children.	Can be self-insufficient: very small person needs special adaptations

(Contd.)

(Contd.)

• Extreme in Weight	Extreme obesity	Usually self-insufficient, need special furniture, furnishings, should be protected from excessive exertion, e.g. climbing stairs.
• Extreme in Size	Dwarfism, gigantism, small children.	Can be self-insufficient: very small person need special adaptations.

Some general design considerations in interiors *(Bhandari, 1992)* findings were as under :

Access and circulation

Facilities (including emergency exits) should be accessible for persons with disabilities and elderly people. Resting space in between floor climbing should be provided. Corridors should be wide with slip resistant floor coverings and highlighting without glare.

Handrails

Round-section handrails should be provided on both sides and allow for a wheelchair (distance 86 - 96 cm minimum). The height of the handrails is to be 90 cm with a good grip to support the weight of an average adult. The ends of the handrails should turn into walls to indicate they are ending.

Floor coverings

Carpet, tiles or plastic are recommended as finish.

Doors

Entrances and doors should be under cover and easily accessible for wheelchairs. Doors should be easy to open, i.e. automatic sliding or conventional doors with low effort door knobs. A shelf beside doors helps old people to put down packages while opening doors.

Elevators

They are essential, for all buildings higher than two storeys. It encourages mobility for those requiring canes, wheelchairs or crutches.

The elevator design should allow enough time for people to leave and get into the elevator. Elevators should adjust to floor levels at every stop and handrails should be provided on three sides of it. Doors should open and close slowly.

Stairs and ramps

Stairs should not have more than 10 steps at a stretch. They should not have projected risings or open steps. Finish should be non-slippery. Handrails should be on both sides and extending at least 30 cm beyond the top and bottom of stairways and ramps. Ramps should be introduced where changes in level occur. It is more economical than an elevator. To prevent wheelchair users from rolling off, the edges on both sides should have a curb at least 5 cm high or a guard rail 20 cm above the ramps. A gradient of 1:20 is preferred but must never increase 1:12 and for not more than 9 metres distance. Ramp landings should extend beyond door swings by at least 90 cm. Top and bottom landings should be at least 150 cm long whereas the intermediates ones 170 cm and spaced at not exceeding 9 metre intervals or where a change of direction occurs. Handrails should be provided on both sides at a height of 85 cm. The minimum clearance between handrails for a wheelchair should range between 86 cm to 96 cm. Round section handrails 4 cm in diameter are recommended to assure a firm grip.

Fire safety

High-rises should be equipped with common access and egress corridors with staff assistance. It is recommended to have fire-separated areas of refuge. Those with sight, hearing or mobility disabilities might need special assistance in case of fire.

Security, safety and privacy

An emergency system of battery lighting should be provided throughout buildings. People need to feel secure but not lonely. The alarm system should be centrally controlled in the building or close enough so that help can arrive fast. It should be located strategically, one in the bathroom — a site of many accidents — and one in the bedroom.

Thus, if a house is designed (interior, furniture and equipment) according to the requirements of elderly people, *Benktzon (1993)* agreed that this same house can as well be more than appropriate for the other age groups. Design for old people will move "from the margins towards the mainstream of product development in the future".

Conclusion

The survey of literature is a vital part of a research endeavour. From the review of existing literature of various topics, namely dimensions of old age homes, infrastructure facilities, housing satisfaction, health hazard/health problem and guidelines for functional constructive unit for old age homes following conclusions were drawn:

1. Mostly research studies in the field of ergonomics and housing were conducted in abroad. Hardly any review was found related to ergonomic assessment of old age home under Indian conditions.
2. There are few data on the health status of elderly people who were the inmates of old age homes despite the increase in longevity and the importance of maintenance of autonomy.
3. Anthropometric data of the elderly are only available for a few countries, e.g. the USA. Even when static anthropometric data are available, there are many problems with the functional dynamic anthropometry of the elderly. Apart from a few body dimensions such as height and

body weight, there is practically no comprehensive anthropometric information on other body dimensions. Anthropometry has remained relatively untouched. At present there are no published data on the anthropometry of elderly people in India.

The main objective was to fill in the gap in information on anthropometric measurements needed to design old age homes ergonomically and related living facilities for elderly people in India.

3

Materials and Methods

The methods and procedures applied in the selection of sample, identification of variables, their measurement and the statistical tools used in analyzing the data form an important part of any research venture. In order to facilitate systematic presentation, it is divided into following sections :

3.1 Research design
3.2 Conceptual framework
3.3 Selection of variables
3.4 Operational definitions
3.5 Selection, construction and description of tool
3.6 Selection of sample
3.7 Data collection procedure
3.8 Data processing and analysis

3.1 Research design

Research design is the arrangement of conditions for collection and analysis of data in a manner that aims to combine reference to the research purpose with economy in procedure. Research design is a plan that specifies the sources and types of information relevant to the research problem, approach for gathering and analysis of data. It also includes time and cost budget as most of the studies are conducted under these two constraints. In a study, research design is needed because it facilitates the smooth sailing of the various research operations in a systematic manner. In the present study descriptive research designs were chosen to find out existing infrastructure facilities, dimensions of the old age home, dimensions of furniture, ergonomic assessment of old age homes and related perceived pain. On the other hand, experimental research designs were planned to find out anthropometrical measurements of elderly people, postural stress while storing personal belonging to storage areas, environmental parameters and health hazards caused.

3.2 Conceptual framework

Conceptual framework of the study represents the relation of the variables with respect to objectives of the study. It is proposed that if the old age homes are constructed on the guidance of ergonomic principles and according to the recommendations given in National Building Codes of India, it increases satisfaction among elderly people and reduces health hazards among them. The above statement is the function of anthropometrical measurement of elderly people, their personal variable and the dimensions of old age home, furniture and furnishing (Fig. 3.1).

3.3 Selection of variables

A variable, as the name implies, is something which varies. In other words, variables are the characteristics of conditions that are manipulated, controlled or observed by the experimenter. Variables under study are categorized into two groups, *viz*. dependent variables and independent variables. The schematic representation of the variables used in the study is given in Fig. 3.2.

3.3.1 Dependent variables

Dependent variables are measures of the change imposed by the independent variables. Following dependent variables were identified in the study :

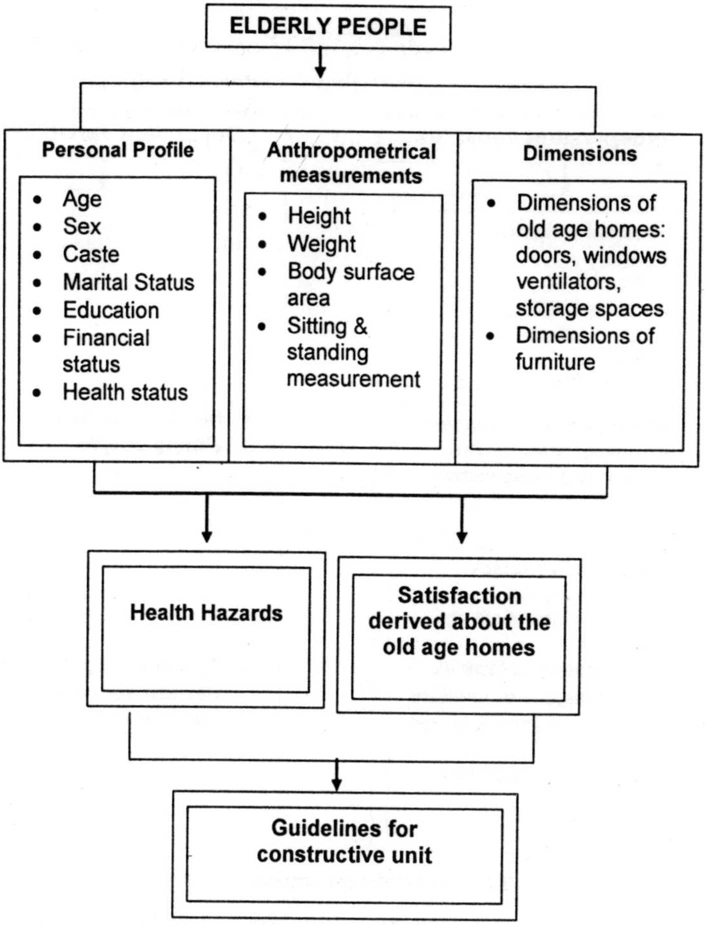

Fig. 3.1 : Conceptual framework of the study.

1. Health hazards on elderly people, and
2. Satisfaction derived from the old age home.

3.3.2 Independent variables

A variable that is antecedent to the dependent variable is termed as independent variable or in other words those variables, which are

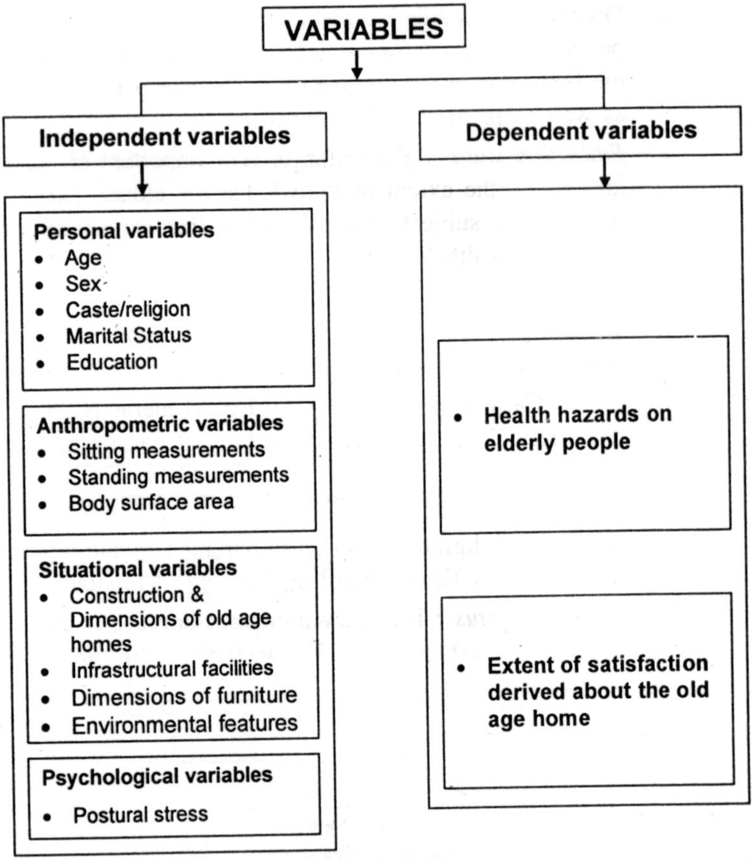

Fig. 3.2 : Schematic presentation of the variables

predicted from personal and situational variables. Independent variables relevant to the present study were grouped under personal variables.

3.3.2.1 Personal variables

1. *Age :* Age was measured in terms of completed years of age at the time of study. Age of the elderly people was assumed to have an impact on health hazard and extent of satisfaction derived.

2. *Sex :* Sex was studied as number of male and female elderly people considered for the study. There are differences in the body dimensions of male and female pelvic structure as well as in the body composition.
3. *Education status :* Education is very important because it may affect the extent of knowledge the elderly people is having on a subject or topic and in this way it can also affect the health hazards.
4. *Marital status :* It is also important variable which shows the family structure of elderly people and types of inmates living in old age homes.
5. *Caste :* Caste refers to a class or distinct hierarchical order of society. The operational measure of caste was taken from socio-economic status scale developed by *Shah B. (1987).*
6. *Religion :* Religion is also an important variable which is categorized as Hindu, Muslim, Sikh and Christian.
7. *Financial status :* It is very important because it can also affect the health and satisfaction derived by the elderly people.

3.3.2.2 Anthropometrical variables

Anthropometrical measurements are often used in ergonomic design. Design development for comfortable working and seating involves anatomical, physiological and biomechanical considerations of the human body. Anthropometrical measurements have direct relation with the suitability and fitness of respondents to the furniture.

3.3.2.3 Situational variables

1. *Construction and dimensions of old age homes :* These are important variables which facilitate to assist the homes ergonomically like size of the habitat room, length and width of storage spaces etc.
2. *Infrastructure facilities :* In the old age homes these are important elements that can affect the health of the

individuals and the comfort level either favourably or adversely.

3. *Dimensions of furniture (table, chair, cot) :* Discomfort and bad postures associated with badly designed chairs, tables, and cots are factors which may affect the health of the elderly people, therefore, seat height, depth, and chair back design become important aspect.

4. *Environmental features of old age homes :*

   1. *Illumination in the rooms :* In the aging eye, the cornea scatters light before it reaches the retina, the lens absorbs more light, and pupil size is reduced, allowing less light to reach the retina *(B. Schneider and Pichora-Fuller, 2000).*
   2. *Ambient temperature :* Extremes of cold and heat can obviously cause adverse health outcome. Housing that protects against these extremes is essential to health. Rapid changes in temperature can also be a cause of deterioration to health and should be guarded against.
   3. *Relative humidity :* High or low humidity than the required level also makes the environment less efficient for work. The relative humidity should not normally exceed 70 per cent for comfortable living.
   4. *Noise level :* Noise can be described as unwanted sound. It has been long recognized that individuals who have been exposed to high noise levels over a long period of time may suffer a hearing loss. Measurement and quantification of noise level provide the fundamental data for assessing hearing exposure risk.

3.3.2.4 Psychological variables

1. *Postural stress :* Postural stress may increase with age *(Borger et. al. 1999).* Incorrect posture produces both physical and mental fatigue. According to *Woodson (1981)* least fatigue occurs when body can be kept in balance that

is when there is the least demand on muscular system to keep the body upright.

2. *Extent of satisfaction derived :* A three-point rating satisfaction scale was developed to measure the extent of satisfaction derived by the elderly people from the old age homes. It was judged for content validity and reliability using split-half method.

3.4 Operational definitions

Certain concepts were operationally defined for the measurement of variables in the present study. These are as under:

Ergonomics

Ergonomics is concerned with the designs of system in which people carry out work. It is the science of fitting jobs to the people. A large number of factors play a role in ergonomics; these include body posture and movement (sitting, standing, lifting, pulling and pushing), environmental factors (noise, vibration, illumination, climate,) chemicals and operation as well as tasks and jobs. These factors determine to a large extent safety, heath, comfort and efficient performances at work and in everyday life.

Ergonomic assessment of old age home

It deals with the assessment of old age home on the basis of ergonomic principles and National Building Codes of India.

National building code

These are standards, which provide guidelines for the planning and general building requirements.

Anthropometrical measurement

It is concerned with measurement of human body, from anatomy point of view. Dimensions of body providing useful data to architect for designing the work surface, storage space etc.

Body surface area

Nomogram (Fig. 3.3) was used for body surface area from weight, and standing height according to the formula of *Du Bois and Du Boiis (1916)* modified by *Banerjee and Sen (1955)*. In order to find the body surface area of elderly people, height in column (i) was correlated with the weight values tabulated in column (ii) by extrapolating the two columns with the help of ruler, column (iii) having body surface area values was placed between column (i) and column (ii), while this extrapolation the values in column (iii) touching the edge of the ruler indicates the body surface area (in sq. metres).

1. *Eye height (Inner canthus height, standing) :* Vertical distance from the floor to the inner corner of the eye, subject stands erect, looking straight ahead (Fig. 3.4).
2. *Shoulder height (standing) :* Vertical distance from the floor to the most lateral point on the superior surface of the acromion process of the scapula, while standing erect (Fig. 3.4).
3. *Elbow height (standing) :* Vertical distance from the floor to radiate the depression at the elbow between the bones of the upper arm (humerous) and forearm (radius), subject stands erect, arm hanging naturally at sides (Fig. 3.4).
4. *Sitting height :* Vertical distance from the sitting surface to the top of the head, subject sits erect, looking straight ahead, with knees and ankles forming right angles.
5. *Eye height (Internal canthus height, sitting) :* Vertical distance from the sitting surface to the inner corner of the eye, subject sits erect, looking straight ahead (Fig. 3.5).
6. *Shoulder height (Acromion height, sitting) :* Vertical distance from the sitting surface to the most lateral point on the superior surface of the acromion process of the scapula, while sitting erect (Fig. 3.5)

Nomogram for Body Surace Area

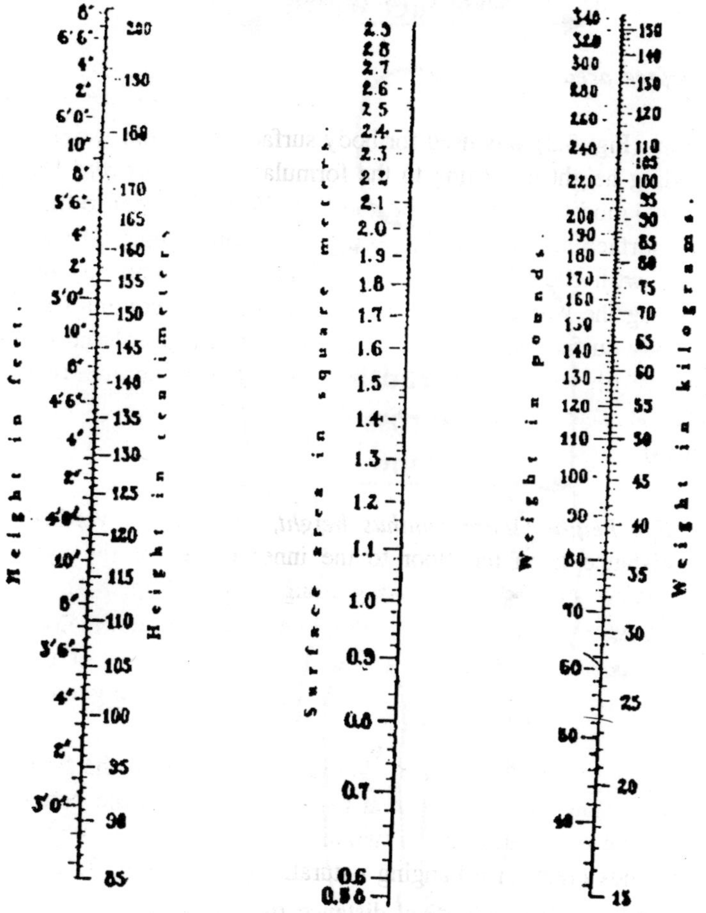

Nomogram for surface area from weight and standing height according to the formula of Du Bois ul Du Bois modified by Banerjee and Sen. A line joining the height of an individual crosses the ule of surface area at the required value.

REFERENCE.

Banerjee, S., and Sen. R. N. (1955) ... *Jour. Appl. Physiol.*, 7, p. 585

Du Bois, D., and Du Bois, E. F. (1916) ... *Arch. Int. Med.*, 17, p. 863.

Feldman, W. M. (1935) — 'Biomathematics', 2nd ed., Charles Griffin & Co., Ltd., London, p. 146.

Feldman, W. M., and Umanski, A.J.V. (1922) *Lancet*, 1, p. 273.

Fig. 3.3 : Nomogram for calculating body surface area

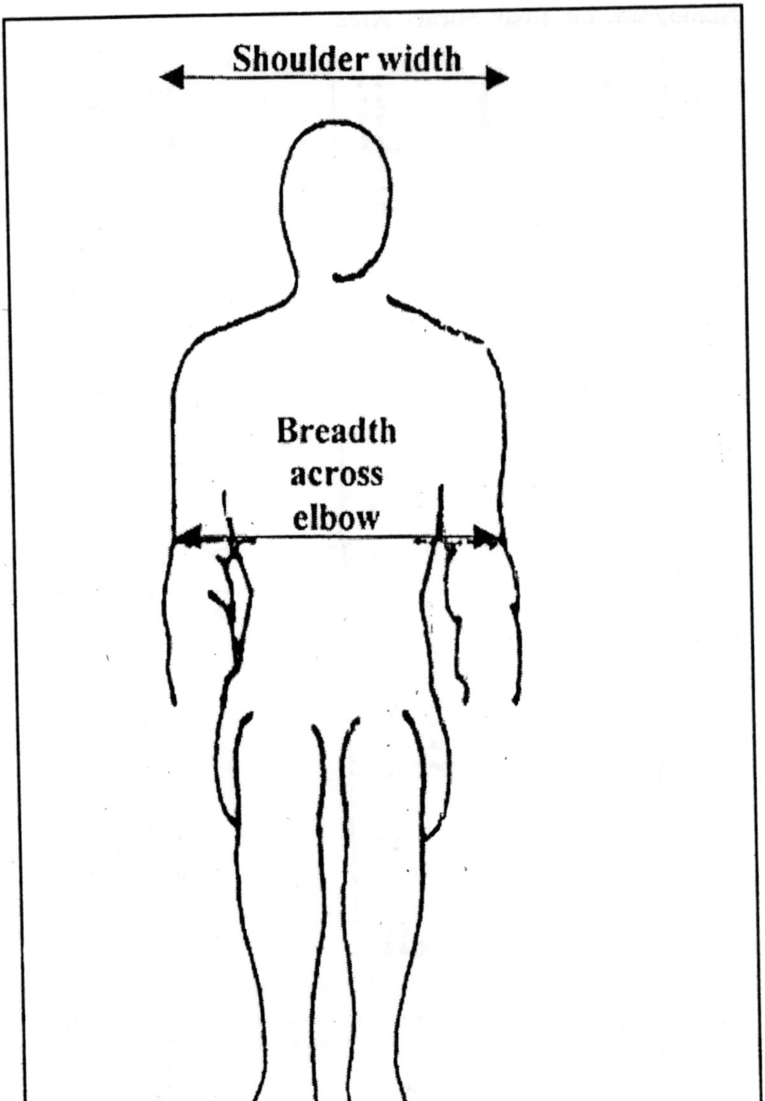
Shoulder width
Breadth across elbow
Breadth across buttocks

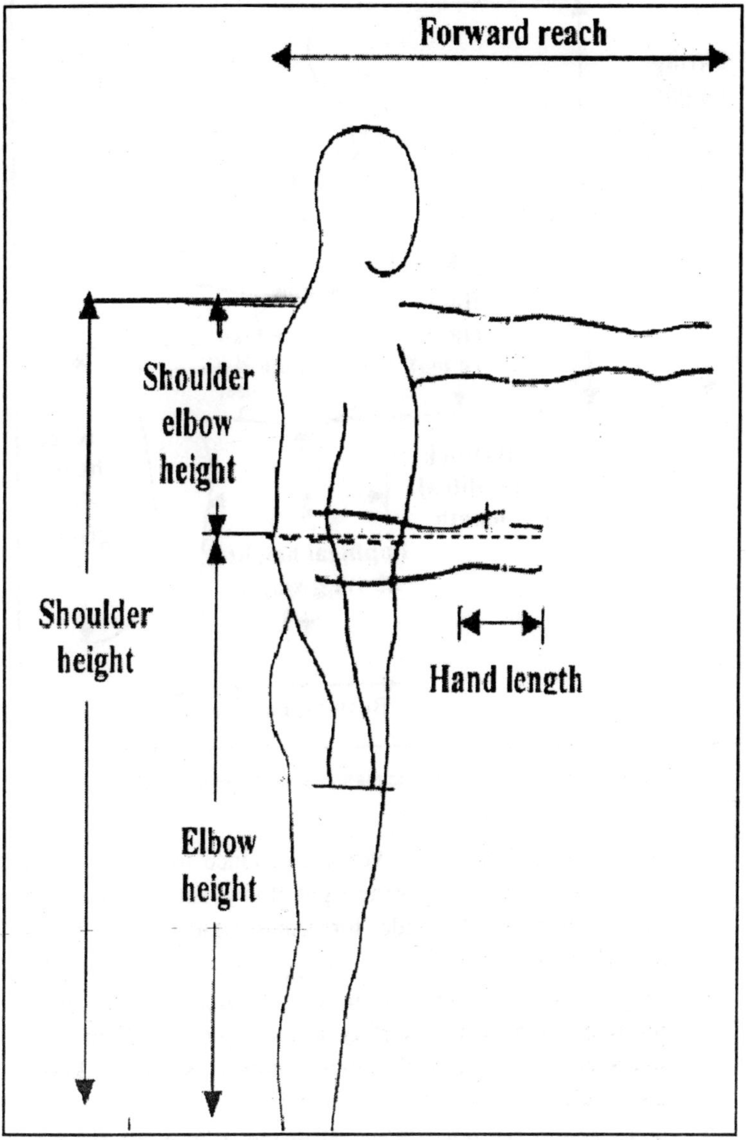

Fig. 3.4 : Static anthropometric measurements (standing posture)

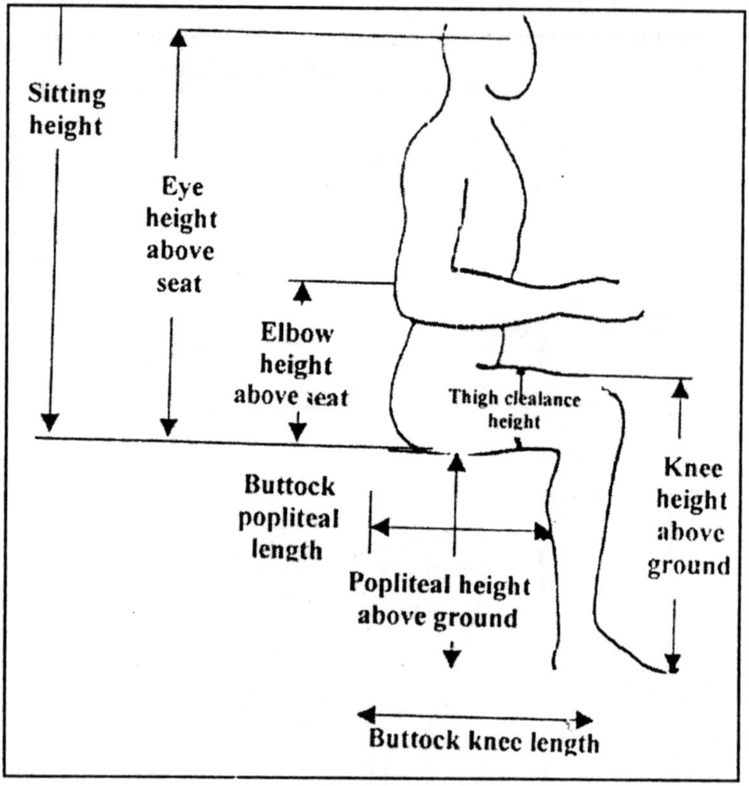

Fig. 3.5 : Static anthropometric measurements (sitting posture)

7. *Elbow height (sitting) :* Vertical distance from the sitting surface to the bottom of the right elbow, subject sits erect, upper arm vertical at side, forearm at right angle to upper arm (Fig. 3.5).
8. *Thigh clearance height (sitting) :* Vertical distance from the floor to the sitting surface to the top of the thigh at its intersection with the abdomen, subject sits erect, knees and ankles at right angles (Fig.3.5).
9. *Knee height (sitting) :* Vertical distance from the floor to the uppermost point on the knee, subject sits erect, knees and ankles at right angles (Fig. 3.5).
10. *Popliteal height (sitting) :* Vertical distance from the floor

to the underside of the thigh immediately behind the knee, subject sits erect, knees and ankles at right angles, bottom of thighs and back of knees barely touching the sitting surface (Fig. 3.5).

11. *Buttock knee length :* Horizontal distance from the plane of the near most point on the buttocks to the front of the knee, subject sits erect, knees and ankles at right angles (Fig. 3.5).
12. *Shoulder-elbow length :* Distance from the top of the acromion process (at the uppermost edge of the shoulder) to the bottom of the elbow, subject sits erect, upper arm vertical at side and making a right angle with the forearm.
13. *Forearm hand length :* Distance from tip of elbow to tip of the longest finger, while sitting erect, upper arm vertical at side, forearm, hand and fingers extended horizontally.
14. *Elbow to elbow breadth :* Maximum horizontal distance across the lateral surface of the elbows, subject sits erect, upper arms vertical and touching the sides, forearms extended horizontally.
15. *Hip breadth (sitting) :* Maximal horizontal distance across the hips, subject sits erect, knees and ankles supported at right angles, knees and heels together.
16. *Buttock-poptileal length :* It is the distance from the buttocks to the back of the knee (Fig.3.5).
17. *Standing vertical reach :* Highest vertical reach used to determine maximum allowable height for overhead controls so that they are reachable by the shortest users [Fig. 3.4 (b)].

Ambient temperature

It is the degree of hotness or coldness in the environment. Comfortable body temperature is 20°C and 23°C. Thermo hygrometer is used for measuring indoor humidity and temperature. The upper display shows indoor temperature while lower display shows humidity level. The user can view maximum and minimum temperature/humidity records by pressing maximum and minimum key.

Sound level

It is the ability of respondent to hear clearly, without stress and strain to ears. Sound is measured in decibel; with the help of sound level meter.

Illumination

It is defined as means of providing light. Unit of light is Lux, Lux-meter is used to measure the intensity of light available in the room.

Stress

The summation of stimuli that tend to upset the physiological equilibrium of an individual. The degree of stress is measured by the degree of departure of the equilibrium state from the normal. For example, heart rates increase during heat exposure.

Rating of intensity of pain

The selected elderly people were asked to answer a pain questionnaire and to stand by a plumb line for pain perceived. Response on each body part in the scale was scored from 0-10 which was interpreted as in the figure (Appendix-I) like 0 no pain, 0.5 extremely weak pain, 1 very weak, 2 weak, 3 moderate, 4 somewhat strong, 5 strong, 6 in between strong and very strong, 7 very strong, 8 and 9 in between very strong and extremely strong pain and 10 extremely strong and the black dot shows maximal pain which is beyond 10 points.

Goneometer (Dualer IQ)

The Dualer IQ measure spine ranges of motion using dual

inclinometers. The primary and secondary sensors are only connected when used for dual inclinometer measurements, such as spine Range of Motion (ROM). In dynamic mode, it automatically records the furthest end point (greatest range of motion) for each movement.

Gross spinal flexion

Place both Primary and Secondary sensors on a wall and press the start/stop button to establish 0. With the subject standing upright in a neutral position, knees straight and weight equally on both feet. Press the Dynamic Static button to display the Dynamic/Static mode. Place the Primary sensor at T1 and the Secondary sensor over the sacral mid-point in the sagittal plane and have the subject flex forward maximally record angle. Dualer IQ simultaneously measures and display provides T1 ROM (Primary sensor), Sacral ROM (Secondary sensor) and Total Range of Motion (RoM) with each flexion. The Total RoM value is the difference in angle between the Primary and Secondary inclinometers. It is produced by subtracting the secondary movement value from the primary movement value (Fig. 3.6).

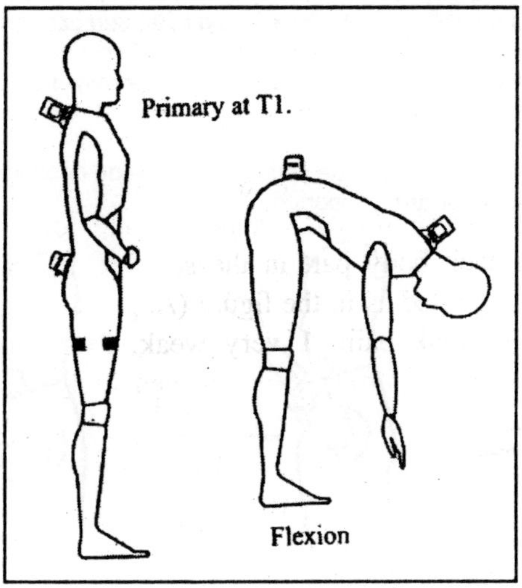

Fig. 3.6 : Gross spinal flexion

Shoulder flexion

Single inclinometer is used in Dynamic/Static mode. Turn the primary sensor on position. Press the Dynamic/Static button to display the static mode (symbol). Place the inclinometer on a flat surface, such as a wall, or on the subject and press the start/stop button to establish 0. Attach the Primary sensor to the upper arm with a strap. Have subject stand with arm down and shoulder in neutral position. Have subject flex the shoulder maximally; press the start/stop button to record the angle (Fig. 3.7).

Shoulder external rotation

Attach the Primary sensor to the forearm with a strap. Have subject stand or sit with shoulder abducted to 90 degrees, elbow flexed to 90 degrees, and upper arm in neutral rotation. Have subject externally and internally rotate the shoulder (Fig. 3.7).

Shoulder Internal/External Rotation
Single inclinometry in Dynamic Static mode

Attach the Primary sensor to the foream with a strap. Have subject stand or sit with shoulder abducted to elbow flexed to 90 degrees, and upper arm in neutral rotiotion. If torso sway is a problem, the Secondary so attached and positioned over the sternum. Have subject externally and internally rotate the shoulder.

Note : This test can also be performed in aprone or supine position with the upper arm supported.

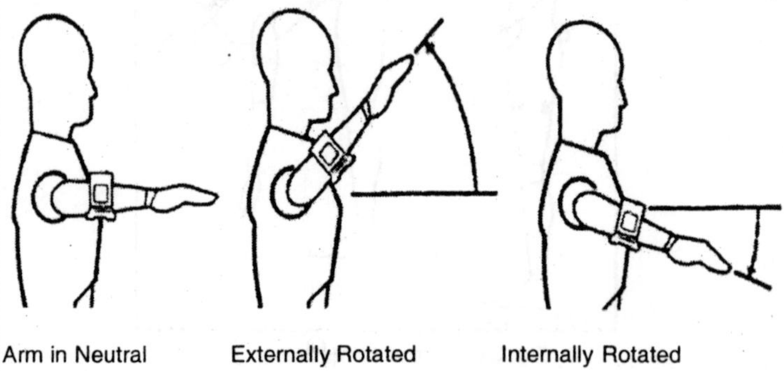

Shoulder Flexion/Extension
Single inclinometry in Dynamic or Static Mode

Attach the Primary sensor to the upper arm with a strap. Have subject standing with arm shoulder down and extended. Have subject flex and extend the shoulder maximally.

Note : For motion between the humerous and scapula, attach the Secondary sensor and hold it firmly against during motion. If torso sway affects the test, attach and hold the Secondary unit against the sternum.

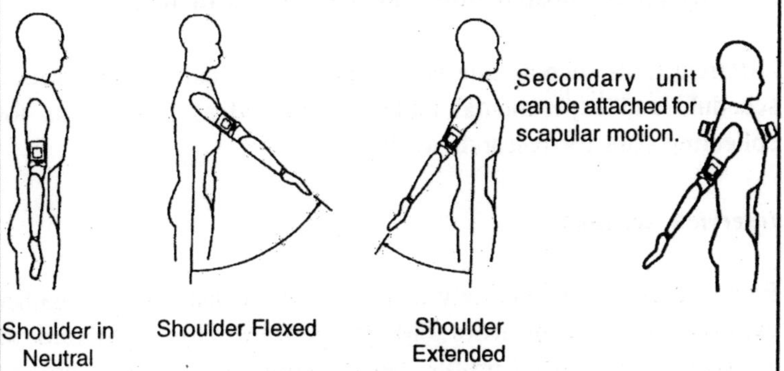

Fig. 3.7 : Shoulder flexion/external rotation

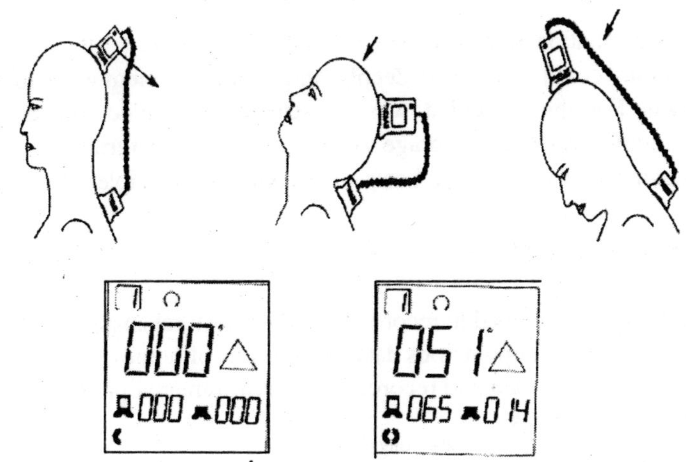

Fig. 3.8 : Cervical flexion/extension.

Cervical flexion/extension

For taking measurement with the subject seated and cervical spine in neutral, place the Seconday sensor in the sagittal plane at T1 and Primary sensor in the sagittal plane on the calvarium. Make sure the sensors do not slip during flexion and extension. If slippage occurs the Primary sensor can be positioned above the ear using strap (Fig.3.8), now subject flex and extend cervical spine.

3.5 Selection, construction and description of tool

Different tools such as interview schedule, checklists, observation schedule and experimental tables were used as instruments for collecting data for research work.

Interview schedule

Precoded personal interview schedule was constructed to gather general information from elderly people. This consisted of information on general profile, health status and health hazards of elderly people.

Checklist

The information was collected through observation so that whole questions could be probed deeply and a holistic picture could be drawn. For this checklist was constructed to collect the depth information related to storage areas, furniture and furnishing and assessment of risk on the basis of ergonomic principles.

Experimental table

Tables were developed to record the data of postural stress of elderly people, their anthropometrics data; dimensions of homes, dimensions of furniture provided and recording of environmental parameters of the old age homes.

Development of satisfaction scale

A three-point rating satisfaction scale was developed to measure the extent of satisfaction derived by the elderly people from the old age home.

1. *Item collection :* A number of items which could contribute towards the satisfaction derived from different housing features by the elderly people were collected from the literature and concerned experts in the field. An item relevant to study, on existing conditions of the old age home and infrastructural facilities available was collected.

   Following criteria were considered while selecting items for the scale :

   - The statement should be as simple as possible.
   - They should be clear, brief and straightforward.
   - Double negatives should not be present in any statement.
   - They should have a certain difficult value and that should promote thinking of the respondents.

   These items were then thoroughly screened and cited in order to make them more simple, clear and meaningful to elicit maximum information.

2. *Establishment of content validity of tool :* Validity is the most critical criterion and indicates the degree to which an instrument measures what it is supposed to measure. In other words, validity is the extent to which differences found with a measuring instrument reflect true differences among those being tested.

   Content validity is the extent to which a measuring instrument provides adequate coverage of the topic under study. It was determined by using a panel of persons, which judges how well the measuring instrument meets the standards. For content validity of the satisfaction scale ten experts were selected from the College of Agriculture, and College of Home Science, Govind Ballabh Pant University of Agriculture and Technology, Pantnagar. These experts were requested to judge the content validity of the scale. For this a score sheet was prepared for the judges and the screening of this was done on the basis of following criteria :

- Those statements, selected by 70 per cent or more judges on relevance were retained in the scale.
- Those statements, which were not cleared by 70 per cent or more judges were recasted to make them clear and meaningful.

3. *Establishment of reliability :* To establish the reliability of the scale, pretesting on a sample of 33 elderly people were taken. The responses of elderly people on each statements of satisfaction scale was quantified by assigning score, where score 1 was assigned to response "Not satisfactory", score 2 was assigned to response "satisfactory" score 3 was assigned to response "highly satisfactory".

Establishment of reliability of the scale

Reliability is the level of internal consistency or stability of the measuring devices. The reliability, therefore, refers to the test consistency.

The reliability of the scale was tested with the application of Split-Half Technique. The zero order correlation between even and odd numbered items was worked out as a measure of reliability. The coefficient of reliability was computed with the help of Spearmen Brown Prophecy formula:

$$r = \frac{\Sigma(xy)}{\sqrt{\Sigma x^2 \times \Sigma y^2}} \tag{3.1}$$

$$rtt = \frac{2rhh(xy)}{1 + rhh}$$

Where,

r = Correlation co-efficient

rtt = Reliability of the total test estimated

rhh = Correlation between halves

The correlation coefficient and reliability coefficient computed values were 0.66 and 0.79, respectively, thereby revealing commendable reliability of the scale.

Testing of flooring material

Two types of flooring materials, *i.e.* cement and concrete and marble stone were generally found in selected old age homes. The flooring materials were collected from supplier and were tested for strength and water absorption in the laboratory of Department of Civil Engineering at College of Technology of G.B. Pant University of Agriculture and Technology, Pantnagar (U.S. Nagar), Uttarakhand and then compared with the Indian standards codes.

Ergonomically designing of chair

For designing the chair ergonomically, the anthropometric data in sitting position were collected. The chair was designed by Computer Application in Designing at College of Technology, Department of Mechanical Engineering, G.B. Pant University of Agriculture and Technology, Pantnagar (U.S. Nagar), Uttarakhand according to the inputs received from anthropometric data and Indian Standard Codes.

3.6 Selection of sample

3.6.1 Selection of the locale

The present study was conducted in two regions of Uttarakhand State, *i.e.* Kumaon Haldwani block of Nainital district and Garhwal (Haridwar and Dehradun districts).

3.6.2 Sampling design

Purposive sampling design was used to select the study area and samples on the basis that not much research work has been carried out in these areas.

3.6.3 Sample size

A sample size is a definite plan for obtaining a sample from a given population keeping in view nature of universe, nature of study, type of sampling technique, availability of finance, size of population, time available, etc. Fifteen elderly people from each selected old age homes Nirmala (Haldwani block); Geeta Kutir and Vridh Sewa (Haridwar), and Prem Dham (Dehradun) were selected randomly for the study, thus making total sample size of sixty elderly. Elderly people who were unable to stand unassisted were excluded from the study. For the collection of experimental data, twenty per cent of selected sample size of sixty elderly peoples was selected. The unit of enquiry comprised elderly people of the old age homes. Schematic representation of sampling design is presented in Fig.3.9.

3.7 Data collection procedure

The data for the present study is collected in two different phases, one following another.

Descriptive data

The data was collected in two phases, in first phase descriptive data were collected from sixty elderly people personally by using pre-coded interview schedule. A visit was made to old age homes prior to the data collection in order to take permissions from the concerned authorities. The investigator explained the purpose of investigation to the elderly people and conducted interview between 10-12 A.M. during the months of August-October 2006.

Prior to this a pilot study was conducted to test the

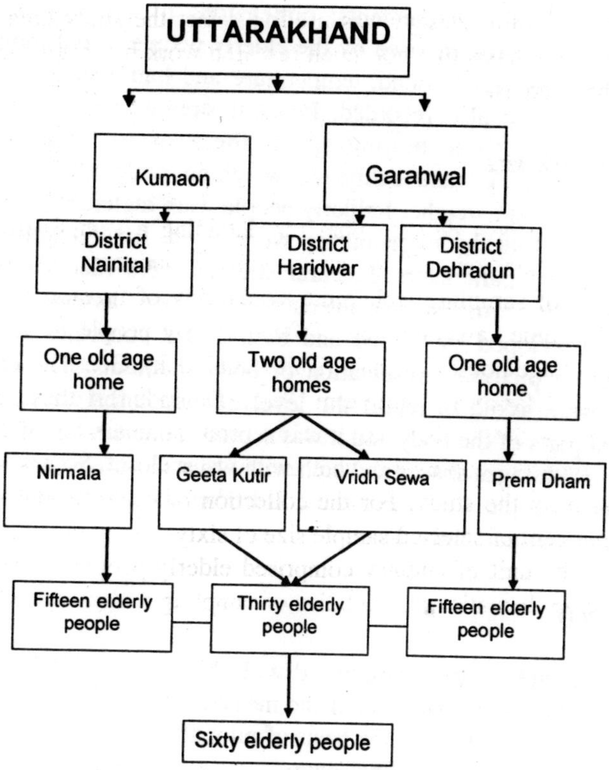

Fig. 3.9 : Schematic representation of sampling design.

appropriateness of the instruments developed for the study. The observation method was also used for collection of information about the existing condition and infrastructure facilities provided in the old age homes. Checklists were developed for ergonomic assessment of old age homes and assessment of risk. Besides this satisfaction scale was developed and used for collection of the data regarding extent of satisfaction level derived by the elderly people.

Experimental data

In second phase the experimental data was recorded from twenty per cent selected samples from the total samples. Experimental tables were made for the collection of data regarding the anthropometrics measurements; dimensions of old age homes, dimensions of furniture used by the elderly people. Environmental parameters like light, sound, temperature and humidity in the old age homes were also recorded. Postural stresses were measured while storing their belonging in the storage spaces. An anthropometrics kit and calibrated weighing scale were used to record height and weight of elderly people. Lux-meter, Sound level meter, and thermo hygrometer were used for recording of environmental parameters. Postural stresses were measured with the use of Goneometer. Pain intensity levels were assessed with a method developed by Corlett and Bishop that combines a rating scale and a pictorial presentation of body diagrams. The elderly people were asked to report the level of discomfort they feel in different parts of the body and it was scored for, chair, cot, window, ventilator, storages spaces and bath with water closet. Scores earned by elderly people were interpreted as higher the scores, higher the pain in particular body part.

Pilot study

The pilot study was accomplished with thirty-three elderly people to check the appropriateness of the measuring tool. Minor changes were made in the tool on the basis of pilot study and then the same was finalized to collect final data.

3.8 Data processing and analysis

The collected data were coded and transferred into coding sheets which were than tabulated and transferred to comprehension tables and graphs to give a clear picture of the findings. Frequencies, percentages, means, standard deviation and range were computed for relevant data.

3.8.1 Categorization of the variables for the purpose of analysis and tabulation

Categorization is the first step in any analysis categories are formed according to research problem and purpose. For the present study variables were categorized in the following manner:

1. *Age :* 60 - 70 years
   70 - 80 years
   80 + years
2. *Sex :* Male
   Female
3. *Assessment of perceived risk :* The scores earned by elderly people were categorized in three ranges *i.e.*

Score	*Category*
40-50	Low
51-61	Moderate
62-72	High

4. *Extent of satisfaction derived :* The responses of elderly people on each statements of satisfaction scale were quantified by assigning scores.

Statements	*Response*	*Score*
	Not satisfactory	1
	Satisfactory	2
	Highly satisfactory	3

The scores earned by elderly people were categorized in three ranges *i.e.*

Score	*Category*
40-50	Low
51-61	Moderate
62-72	High

3.8.2 Coding and tabulation

The information collected was coded into the coding sheets and was then transferred into tabular form :

1. General profile of the elderly people.
2. Health status of the elderly people.
3. Age and sex-wise distribution of occurrence of an accident.
4. Dimensions of old age homes, storage spaces, doors, windows, and ventilators.
5. Descriptions of infrastructure facilities available in old age homes.
6. Dimensions of furniture.
7. Anthropometric measurements of the elderly people.
8. Descriptions of environmental parameters in the old age homes.
9. Postural stresses.
10. Assessment of pain perceived by the elderly people.
11. Assessment of risk.
12. Ergonomic assessments of old age homes.
13. Extent of satisfaction derived from old age homes.
14. Testing of hypotheses.

3.8.3 Statistical analysis of data

The data were coded according to code numbers assigned and were analyzed using descriptive as well as inferential statistics.

3.8.3.1 Descriptive statistics

The findings were presented in frequencies, percentages, means, standard deviations and ranges.

Frequency: It is sum of responses (In numbers).

Percentage: For making sample comparison the percentage values were calculated by dividing the frequency of a particular cell by total number of samples, multiplying by 100.

$$P. = n/N \times 100 \quad \text{... (3.2)}$$

Mean : Mean is the most common measure of central tendency and it can be defined as the value which we get by dividing the total of the values of various given items in a series by the total number of items.

$$\text{Mean (X)} = \bar{x}_i / n = \frac{X_1+X_2+........+X_n}{n} \qquad ... (3.3)$$

Where,

$\bar{X}$ = The symbol we use for mean (Pronounced as X bar).

Σ = Symbol for summation.

X_i = Value of the ith item $X_{1,}\ X_2,\ X_3...$

n = Total numbers of items.

Standard deviation: Is the most widely used measure of dispersion of a series and is commonly denoted by the symbol σ (pronounced as sigma). It is defined as the square root of the average of square of deviation.

$$\text{Standard deviation } (\sigma) = \sqrt{\left\{\Sigma\left(x_i - \bar{x}\right)^2\right\} / (n-1)} \qquad ... (3.4)$$

Percentile: For calculation of percentile

5th P = M – 1.65 × S.D 90th P = M +1.28× S.D ... (3.5)

10th P = M – 1.28 × S.D 95th P = M + 1.65 × S.D

25th P = M – 1.96 × S.D 97.5 th P = M + 1.96 × S.D

50th P = M+/- 0 × S.D

Where,

P = Percentile

M = Mean ($\bar{X}$)

S.D. = Standard deviation (σ)

Pheasant (1996)

Scoring : It was done to analyze the responses given against the statements in the "satisfaction scale" the scoring was done on the basis of the scores given to the 3 points.

Highly satisfactory = 3
Satisfactory = 2
Not satisfactory = 1

The mean scores of statements were calculated by this formula:

$$\text{Mean score} = \frac{\text{Total score of responses}}{\text{Number of elderly people}} \quad \text{... (3.6)}$$

3.8.3.2 Hypotheses testing

The following null hypotheses were formulated and tested :

1. The age and sex of elderly people have no relationship with occurrence of accidents.
2. Anthropometrical measurements of elderly people have no relationship with dimensions of (a) storage areas and (b) furniture.
3. Anthropometrical measurements of elderly males have no relationship with anthropometrical measurements of elderly females.
4. Dimensions of old age homes have no relationship with National Building Codes.
5. Dimensions of furniture have no relationship with Indian Standards Codes.

Test of Independence of factors by Chi-square test was used for testing the hypothesis.

H_0 = Two factors are independent of each other
H_1 = Two factors are dependent.

1. *2×2 value of contingency table:* It is a short-cut method of obtaining the calculated value of Chi-square.

Table : 2 x 2 contingency table

Factor B	*Factor A*		*Total*
	A1	*A2*	
B1	a	b	(a+b)
B2	c	d	(c+d)
Total (a+c)	(b+d)	a+b+c+d=n	

The formula for calculating chi-square from observed frequencies a, b, c and d is:

$$\text{Chi-square} = \frac{n\,(ad-bc)}{(a+c)(b+d)(a+b)(c+d)} \qquad \text{... (3.7)}$$

The decision about independence of factors A and B was taken by comparing calculated value of Chi-square with table value of X^2 for d.f. and $\alpha = 0.05$, *i.e.* 3.841.

2. *Yates correction:* If any of the cell frequency in contingency table of order (2 × 2) is less than 5, then we apply the correction, which is suggested, by Yates.

Formula :

$$\text{Chi-square} = \frac{n\,(|ad-bc|-n/2)^2}{(a+c)\ (b+d)\ (a+b)\ (c+d)} \qquad \text{... (3.8)}$$

Here, $|ad-bc|$ means we consider only the absolute value of the difference *(Agarwal 2002)*. For instance (23–75) will be taken as 52 instead of (–52).

3. *Karl Pearson's correlation co-efficient (r):* Is the most widely used for measuring correlation between two variables. If X and Y are two observations then,

$$r = \Sigma xy \,/\, (\sqrt{\Sigma\, x^2 \Sigma y^2}) \qquad \text{... (3.9)}$$

Where,

$$\Sigma xy = [\Sigma xy - \{(\Sigma x) - (\Sigma y)\} / n\}]$$
$$\Sigma x^2 = [\Sigma x^2 - \{(\Sigma x)^2 / n\}]$$
$$\Sigma y^2 = [\Sigma y^2 - \{(\Sigma y)^2 / n\}]$$

The levels of probability to assess the significance of the computed value of *r* were 5 per cent (0.05) and 1 per cent (0.01) level.

4. *Fisher's t-test:* To test the significance difference between two means from two independent samples, two sample Fisher's *t*-test was used *(Snedecor, G.W. and Cochran, W.G. 1967).*

$$t = (\overline{X}_1 - \overline{Y}_1) / \left\{ \sqrt{\sigma^2 (1/n_1 + 1/n_2)} \right\} \qquad \text{... (3.11)}$$

Where,

$$\overline{X} = \Sigma X / n_1 \qquad \sigma^2 = \frac{(n_1 - 1)S_1^2 + (n_2 - 1)}{(n_1 - 1) + (n_2 - 1)}$$

$\overline{Y} = \Sigma Y / n_2$

n_1 = Number of observation in 1st sample

n_2 = Number of observation in 2nd sample

$$\sigma = \frac{\sum(\overline{X}_1 - \overline{Y}_1)^2 + \sum(\overline{X}_2 - \overline{Y}_2)^2 (\overline{X}_1 - \overline{Y}_1)}{(n_1 - 1) + (n_2 - 1)}$$

Degree of freedom = $(n_1 - 1) + (n_2 - 1)$

$$\text{Also} \quad \sigma^2 = \frac{(n_1 - 1)S_1^2 + (n_2 - 1)S_2^2}{(n_1 - 1) + (n_2 - 1)} \qquad \text{... (3.12)}$$

Where, S_1^2 and S_2^2 are the estimated variances from two samples

$$S_1^2 = (1/n_1 - 1) \{\Sigma X_1^2 - (\Sigma X^2 / n_1)\}$$
$$S_2^2 = (1/n_1 - 1) \{\Sigma Y_1^2 - (\Sigma Y^2 / n_1)\}$$

4

Results and Discussion

The facts and findings of the study emerged from statistical treatment of the collected data have been presented in this chapter. These are in concurrence with the said objectives and are described under different sub-heads are as follows :

4.1 General profile of the elderly people
4.2 Health status of the elderly people
4.3 Age and sex-wise distribution of occurrence of accidents
4.4 Dimensions of old age homes, storage spaces, doors, windows, and ventilators
4.5 Description of infrastructure facilities available in old age homes
4.6 Dimensions of furniture
4.7 Anthropometric measurements of the elderly people
4.8 Description of environmental parameters in old age homes
4.9 Postural stress
4.10 Assessment of pain perceived by the elderly people
4.11 Assessment of risk
4.12 Ergonomic assessment of old age homes
4.13 Extent of satisfaction derived from old age homes
4.14 Testing of hypotheses.

4.1 General profile of the elderly people

General profile of the elderly people was obtained through analysis of data collected and the findings are presented in the ensuing paragraphs.

4.1.1 Age of the elderly people

The elderly people were classified by age into three groups, *i.e.* 60-70 years, 70-80 years and 80+ years. The analysis of data pertaining to age, sex, education marital status, caste, religion, and financial status of the elderly people are given in Table 4.1 and Fig.4.1. It was found that in old age home Ist, the age of the elderly people ranged between 60 to 80+years. About 53 per cent of them were belonging to the age group of 60-70 years, 26.6 per cent to 70-80 years, while 20 per cent to 80+ years. The mean age was 68.73 year with 8.57 S.D.

In old age home IInd, 60 per cent of elderly people were belonging to the age group of 60-70 years, 13.3 per cent to 70-80 years, while 26.6 per cent to 80+ years. The mean age was 72.26 year with 11.03 S.D. In old age home IIIrd, 80 per cent of elderly people were belonging to the age group of 70-80 years, 6.6 per cent to 60-70 years, while 13.3 per cent to 80+ years. The mean age was 73.73 year with 3.84 S.D. In old age home IVth, 66.6 per cent of elderly people were belonging to the age group of 70-80 years, 13.3 per cent to 60-70 years, while 20 per cent to 80+ years. The mean age was 75.46 year with 5.28 S.D. About 46.6 per cent of elderly people belonged to 70-80 years, 33.3 per cent to 60-70 years and 29.1 per cent to 80+ years [Fig.4.1 (a)].

4.1.2 Sex of the elderly people

In old age home Ist, the findings showed that the ratio of elderly male was greater (60 per cent) than elderly female (40 per cent). In old age home IInd and IIIrd, the population of elderly male to female was nearly equal 53.3 per cent and 46.6 per cent, respectively. But in old age home IVth, the population of female was greater (86.6 per cent) than male (13.3 per cent). Overall number of elderly female was greater (56.6 per cent) than male (43.3 per cent). The reason behind this was that most were unmarried females while in

Table 4.1 : General profile of elderly people of old age homes (F =15)

Personal characteristics	*Old age home*				*Total F = 60*
	I	*II*	*III*	*IV*	
1. Age (years)					
60-70	8 (53.3)	9 (60)	1 (6.6)	2 (13.3)	20 (33.3)
70-80	4 (26.6)	2 (13.3)	12 (80)	10 (66.6)	28 (46.6)
80+	3 (20)	4 (26.6)	2 (13.3)	3 (20)	12 (20.1)
Mean	68.73	72.26	73.73	75.46	72.5
S.D.	8.57	11.03	3.84	5.28	7.18
2. Sex					
Male	9 (60)	8 (53.3)	7 (46.6)	2 (13.3)	26 (43.3)
Female	6 (40)	7 (46.6)	8 (53.3)	13 (86.6)	34 (56.6)
3. Educational status					
Illiterate	3 (20)	4 (26.6)	—	—	7 (11.6)
Primary	1 (6.6)	3 (20)	—	—	4 (6.6)
Secondary	4 (26.6)	4 (26.6)	3 (20)	—	11 (18.3)
High School	3 (20)	1 (6.6)	4 (26.6)	—	8 (13.3)
Intermediate	2 (13)	1 (6.6)	3 (20)	3 (20)	9 (15)
Graduate	2 (13)	2 (13.3)	5 (33)	12 (80)	21 (35)
4. Marital status					
Married with alive spouse	1(6.6)	2 (13.3)	2 (13.3)	1 (6.6)	6 (10)
Unmarried	3 (20)	—	9 (60)	7 (46.6)	19 (31.6)
Widow	2 (13.3)	6 (40)	1 (6.6)	7 (46.6)	14 (23.3)
Widower	9 (60)	7 (46.6)	3 (20)	2 (13.3)	21 (35.1)
5. Caste					
General	2 (13.3)	9 (60)	2 (13.3)	10 (66.6)	23 (38.3)
Other Backward Castes	10 (66.6)	4 (26.6)	5 (33.5)	1 (6.6)	20 (33.3)
Scheduled Caste	1 (6.6)	1 (6.6)	2 (13.3)	2 (13.3)	6 (10.0)
Scheduled Tribe	2 (13.3)	1 (6.6)	6 (40)	2 (13.3)	11 (18.3)
6. Religion					
Hindu	12 (80)	11 (73.3)	2 (13.3)	2 (13.3)	27(45)
Muslim	—	—	—	—	—
Sikh/Punjabi	1 (6.6)	4 (26.6)	—	11 (73.3)	16 (26.6)
Christian	2 (13.3)	—	13 (86.6)	2 (13.3)	17 (28.3)
7. Financial status					
No income	—	12 (80)	2 (13.3)	4 (26.6)	18 (30)
Retirement pay	15 (100)	3 (20)	13 (86.6)	11 (73.3)	42 (70)

Note : Figures in parentheses represent percentage.

some cases widowhood forced them to live in the old age homes [Fig.4.1 (b)].

4.1.3 Educational status of elderly people

It is pertinent to mention here that majority of the elderly people were observed to be literate, though the proportion of elderly people

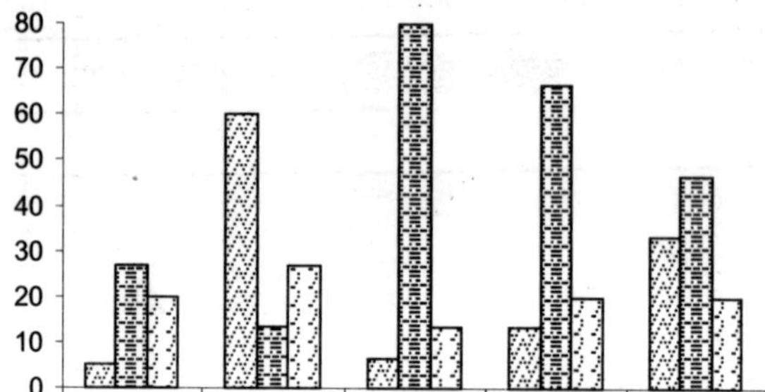

	Old age home I	Old age home II	Old age home III	Old age home IV	Total
60-70 Years	5.33	60	6.6	13.3	33.3
70-80 Years	26.6	13.3	80	66.6	46.6
80+	20	26.6	13.3	20	20

Fig. 4.1(a) : Age-wise distribution of elderly people

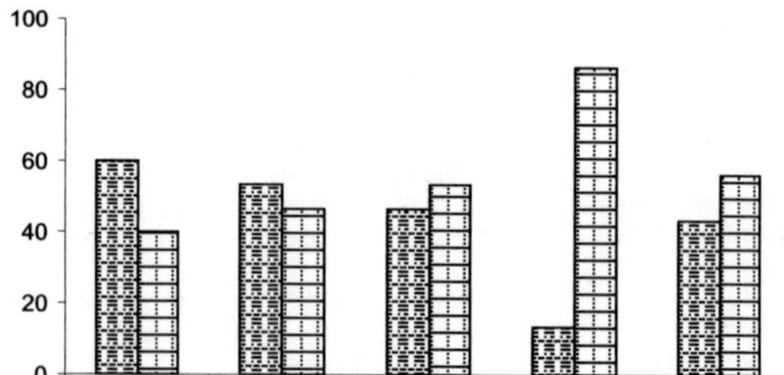

	Old age home I	Old age home II	Old age home III	Old age home IV	Total
Male	60	53.3	46.6	13.3	43.3
Female	40	46.6	53.3	86.6	56.6

Fig. 4.1(b) : Sex-wise distribution of elderly people

by level of education varied, primary, secondary, high school, intermediate, and graduate. A close look at the findings showed that those elderly people in Ist, IIIrd, IVth, *i.e.* paid old age homes were better educated than old age home IInd, while 11.6 per cent illiterate [Fig. 4.1(c)].

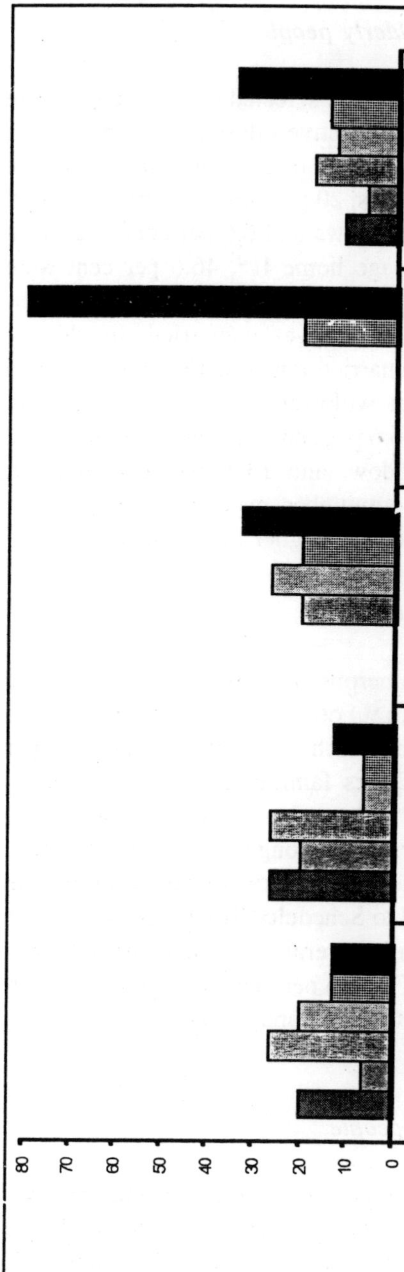

	Old age home I	Old age home II	Old age home III	Old age home IV	Total
Illiterate	20	26.6	0	0	11.6
Primary school	6.6	20	0	0	6.6
Secondary school	26.6	26.6	20	0	18.3
High school	20	6.6	26.6	0	13.3
Intermediate	13.3	6.6	20	20	15
Graduate	13.3	13.3	33.3	80	35

Fig. 4.1(c) : Education-wise distribution of elderly people

4.1.4 Marital status of the elderly people

The background characteristics of the selected elderly people were also studied as it provides the descriptive information, which helps in interpretation of some of the results of the study. In the old age home Ist 60 per cent were widowers, 20 per cent unmarried females, while remaining 13.3 per cent widows and 6.6 per cent lived with their spouse. However, in old age home IInd, 46.6 per cent were widower, 40 per cent widow, while 13.3 per cent lived with their spouse. None of the elderly person was unmarried. In old age home IIIrd, 60 per cent were unmarried females, 13.3 per cent lived with their spouse, 20 per cent widower, and only 6.6 per cent widow. In old age home IVth, 46.6 per cent females were unmarried followed by 33.3 per cent widow, and 13.3 per cent widower. Female widows dramatically outnumber male widowers reasons may be that females survival rate is higher than males.

4.1.5 Caste of the elderly people

The Hindus are divided into numerous castes and sub-castes. There was clear preponderance of (66.6 %) other backward castes followed by general and Scheduled Tribes each 13.3 per cent, and 6.6 per cent belonged to Scheduled Castes families in old age home Ist. However, in old age home IInd, there was dominance of general category (60%) and each 6.6 per cent belonged to SC/ST categories, while 26.6 per cent belonged to other backward castes. In old age home III^{rd,} 40 per cent belonged to Scheduled Tribes category, while only each 13.3 per cent were general, and Scheduled Castes categories. In old age home IVth, 66.6 per cent belonged to general followed by each 13.3 per cent SC/ST, and 6.6 per cent to O.B.C. categories.

4.1.6 Religion of the elderly people

The religions were categorized in four categories, *i.e.* Hindu, Muslim, Sikh/Punjabi, and Christian communities. In old age home Ist, 80 per cent were Hindus followed by Christian and Sikh communities.

In old age home IInd, 73.3 per cent were Hindus followed by Sikh. None of the elderly people belonged to Muslim, and Christian communities. In old age home IIIrd there was preponderance of Christian (86.6%), and remaining 13.3 per cent were from Hindu communities. None of the elderly people belonged to Muslim, and Sikh communities. In old age home IVth, 73.3 per cent were from Sikh/Punjabi communities, and remaining was from Hindu and Christian communities.

4.1.7 Financial status of the elderly people

Financial status affects the economic stability of the elderly people. It was found that 100 per cent elderly people were independent in old age home I[st] as it was paid old age home, and they had either retirement savings or they got from their family members. In old age home IInd, 80 per cent elderly people had no income, and only 20 per cent had their retirement savings. This old age home was free of cost and the authorities provide each and everything needed by them. In old age home IIIrd and IVth, 86.6 per cent, and 73.3 per cent elderly people were independent as they had retirement savings, and only 13.3 per cent, 26.6 per cent had no income, respectively. These three old age homes were paid, and the required fees, *i.e.* Rs. 1,500-2,000 per month was paid by them.

4.2 Health profile of elderly people

Elderly people vary more in health status among themselves than do any other age group. The changes associated with ageing are partly influenced by genetics, race, gender, past medical history, diet and type of hygiene practised throughout the life.

4.2.1 Opinion about health status

The distributions of existing health status, according to elderly people opinions, are outlined in Table 4.2.1. In old age home I[st], about 55.5 per cent males and 33.3 per cent females expressed health status appropriate to their age, whereas 50 per cent females

Table 4.2.1 : Opinion about existing health status of elderly people

Opinion	*Old age home*								*Total F = 60*	
	I		*II*		*III*		*IV*			
	Male F = 9	*Female F = 6*	*Male F = 8*	*Female F = 7*	*Male F = 7*	*Female F = 8*	*Male F = 2*	*Female F = 13*	*Male F = 26*	*Female F = 34*
1 Excellent	1 (11.1)	1 (16.6)	1 (12.5)	1 (14.2)	1 (14.2)	—	1 (50)	—	4 (15)	2 (5.8)
2. Appropriate for age	5 (55.5)	2 (33.3)	2 (25)	1 (14.2)	3 (43)	4 (50)	1 (50)	8 (61.5)	11 (42)	15 (44)
3. Fair enough	2 (44)	—	2 (25)	2 (28.5)	—	—	—	2 (15.3)	4 (15)	4 (12)
4. Poor	1 (11.1)	3 (50)	3 (37.5)	3 (43)	3 (43)	4 (50)	—	3 (23.0)	7 (27)	13 (38.2)
Total	**9 (60)**	**6 (40)**	**8 (53.3)**	**7 (46.6)**	**7 (46.6)**	**8 (53.3)**	**2 (13)**	**13 (86.6)**	**26 (43)**	**34 (56.6)**

Notes : Figures in parentheses represent percentage.
F = Frequency.

and 11 per cent males had poor health status. About 22 per cent males reported fair enough health, while a few had excellent health. None of the females expressed fair enough health. In old age home IInd 25 per cent males and 14.2 per cent females expressed health status appropriate to their age, whereas 37.5 per cent of males and 43 per cent females had poor health status. Twenty five per cent males and 28.5 per cent females exhibited fair health, whereas excellent health was reported by 12.5 per cent males, 14.2 per cent females. In old age home IIIrd, each 50 per cent females and each 43 per cent males expressed their health status poor and appropriate to their age, respectively. None of the elderly people exhibited fair health while a few had excellent health. In old age home IV^{th,} each 50 per cent males were reported their health status appropriate for age and excellent, whereas 61.5 per cent and 25 per cent females exhibited appropriate for age and poor health, respectively. Majority of males and females were of the opinion that their existing health status was appropriate according to age and the elderly males were better than females.

4.2.2 Existing status of sense organs of elderly people

As people start ageing, their strength, endurance, flexibility, vision, gait, posture, hearing, and balance all tend to deteriorate.

4.2.2.1 Eyesight

Difficulty focusing close-up may begin in the 40s, and the ability to distinguish fine details may begin to decline in the 1970s. The distributions of existing status of sense organs of elderly people are outlined in Table 4.2.2. In old age home Ist, 22.9 per cent of males, and only 16.6 per cent females had weak eyesight. Out of them 50 per cent females, and only 22 per cent males had cataract. However, 55.5 per cent males, while a few females used spectacles. In old age home IInd, 12.5 per cent males and 14.2 per cent females had normal eyesight. Twenty-five per cent males and 14.2 per cent females had weak eyesight. Out of them 37.5 per cent males and 14.2 per cent females suffered cataract. Percentages of males and

Table 4.2.2 : Status of sense organ of elderly people of old age homes

Sense organ	Old age home								Total F = 60	
	I		II		III		IV			
	M *F = 9*	*F* *F = 6*	*M* *F = 8*	*F* *F = 7*	*M* *F = 7*	*F* *F = 8*	*M* *F = 2*	*F* *F = 13*	*M* *F = 26*	*F* *F =34*
Eyes										
Normal	—	—	1 (12.5)	1 (14.2)	1 (14.2)	1 (12.5)	—	3 (23)	2 (13)	5 (14.7)
Weak	2 (22.22)	1 (16.6)	2 (25)	1 (14.2)	2 (28.5)	2 (25)	—	2 (15.2)	6 (23)	6 (37.5)
Cataract	2 (22.2)	3 (50)	2 (25)	4 (50)	2 (28.5)	2 (25)	—	3 (23)	6 (23)	12 (35.2)
Use spectacles	5 (55.5)	2 (33.3)	3 (37)	1 (14.2)	2 (28.5)	3 (37.5)	2 (100)	5 (38)	12 (46)	11 (32.3)
Hearing										
Normal	2 (22.2)	—	2 (25)	3 (42.8)	1 (14.2)	3 (37.5)	—	4 (30)	5 (19)	10 (29.4)
Weak	7 (77.7)	6 (100)	6 (75)	4 (57.1)	5 (71.4)	5 (62.5)	2 (100)	9 (69)	20 (76)	24 (70.5)
Use hearing aid	—	—	—	—	1 (14.2)	—	—	—	1 (3.8)	—

(Contd.)

Table 4.2.2 : (Contd.)

Taste										
Normal	8 (88.8)	6 (100)	7 (87)	5 (71.4)	7 (100)	7 (87.5)	2 (100)	11 (84)	24 (92)	29 (85.2)
Weak	1 (11.1)	—	1 (12)	2 (28.5)	—	1 (12.5)	—	2 (15)	2 (7.6)	5 (14.7)
Smell										
Normal	9 (100)	6 (100)	7 (87)	7 (100)	7 (100)	8 (100)	2 (100)	12 (92)	25 (96)	33 (97.0)
Weak	—	—	1 (12)	—	—	—	—	1 (7.6)	1 (3.8)	1 (2.9)
Walking										
Normal	9 (100)	6 (100)	5 (62)	5 ((62)	7 (100)	5 (62.5)	2 (100)	8 (61)	23 (88)	24 (70.5)
Use cane	—	—	3 (37)	2 (28.5)	—	—	—	5 (38)	3 (11)	7 (20.5)
Use Walker	—	—	—	—	—	3 (37.5)	—	—	—	3 (8.8)

Note : Figures in parentheses represent percentage.
M = Male, F = Female, F = Frequency

females using spectacles were 37.5 per cent and 14.2 per cent, respectively. In old age home IIIrd, 14.2 per cent males and 12.5 per cent females had normal eyesight. 28 per cent of males and 25 per cent females had weak eyesight. Cataract was found in 25 per cent females, and 20.5 per cent males. Out of them 37.5 per cent females and only 28.5 per cent males used spectacles. In old age home IVth, 23 per cent, 15.2 per cent, and 23 per cent of females had normal, weak eyesight, and cataract, respectively. Thirty-four per cent of females and almost all the males used spectacles.

Annual eye examinations are helpful not only in ensuring the appropriateness of corrective lenses but also in detecting, in a timely fashion, the many eye disorders.

4.2.2.2 Hearing sense

It becomes more difficulty to hear higher frequencies with age. In old age home Ist, it was found that 100 per cent females had normal sense of hearing as compared to 22.2 per cent males. Males (77.7%) had weak hearing capacity, however, none had hearing aids. In old age home IInd, 42.8 per cent females had better hearing sense as compared to 25 per cent males. 75 per cent of males were weak in hearing as compared to 57.1 per cent females but none had hearing aids to assist them. In old age home IIIrd, 37.5 per cent females were in better hearing condition than 14 per cent males, nearly equal percentage of males and females, *i.e.* 71.4 per cent, 62.5 per cent had weak hearing sense but only 14.2 per cent males used hearing aids. In old age home IVth, 30.7 per cent females had normal hearing sense. Females (69.2%) had weak sense of hearing but none used hearing aids. Hearing declines more quickly in males than females.

Audiometric evaluation should be obtained for persons with hearing impairments to determine possible corrective measures and the benefit of hearing aids.

4.2.2.3 Taste and smell

Taste and smell are two separate and distinct senses that often work together for different purposes. They can detect pleasurable sensations or warn a person of danger. 92 per cent of males were found to have normal taste organs than 85.2 per cent females. However, 97 per cent females had better smell organs than males in all the old age homes.

4.2.2.4 Walking

In old age home I[st] all the elderly people were normal in walking but in old age home IInd, 62.5 per cent males, 28.5 per cent females faced walking difficulties and took the support of cane. In old age home IIIrd, males were better than females, and those 37.5 per cent females faced walking difficulties relied on walkers. In old age home IVth, 61.58 per cent of males and females have no difficulties in walking, while 38.4 per cent females took the support of cane for walking.

4.2.3 Minor and major health problem of elderly people

Old persons with advancing age have to cope with health and associated problems.

4.2.3.1 Minor disease

Minor diseases were identified as dental problem, diarrhoea, constipation, body ache, general weakness, depression and gaseous distension. The distributions of elderly people in old age homes are outlined in Table 4.2.3 exhibited that in old age home I[st], 83.3 per cent and 16.6 per cent females, while 22.2 per cent males suffered from minor illness like dental problem. Constipation, body ache and gaseous distention were more prevalent in males than females, while general weakness and depression was more prevalent in females than males. In old age home IInd, dental problem, diarrhoea, and body ache were more prevalent in females (57.1 per cent, 42.8 per cent, and 100 per cent); while constipation and gaseous distention were more prevalent diseases in males (100%, and 62.5%) than

Table 4.2.3 : Major and minor health problems of elderly people

Health problem	*Old age home*								*Total F = 60*	
	I		*II*		*III*		*IV*			
	M F = 9	*F F = 6*	*M F = 8*	*F F = 7*	*M F = 7*	*F F = 8*	*M F = 2*	*F F = 13*	*M F = 26*	*F F = 34*
Minor diseases										
Dental problem	2 (22.2)	5 (83.3)	3 (37.5)	4 (57.1)	1 (14.2)	4 (50)	—	8 (61.5)	6 (23)	21 (60)
Diarrhoea	—	1 (16.6)	—	3 (42.8)	2 (28.5)	4 (50)	—	5 (38.4)	2 (7.6)	13 (38.2)
Constipation	8 (88.8)	3 (50)	8 (100)	1 (14.2)	6 (85.7)	8 (100)	2 (100)	11 (84.6)	24 (92.3)	23 (67.6)
Body ache	8 (88.8)	2 (33.3)	7 (87.5)	7 (100)	6 (85.7)	7 (87.5)	1 (50)	13 (100)	22 (84.6)	29 (85.2)
General weakness	5 (55.5)	6 (100)	6 (85.7)	7 (87.5)	7 (100)	8 (100)	2 (100)	13 (100)	20 (76.9)	34 (100)
Depression	6 (66.6)	6 (100)	8 (100)	7 (100)	5 (71.4)	8 (100)	2 (100)	13 (100)	21 (80.7)	34 (100)
Gaseous distention	7 (77.7)	3 (50)	5 (62.5)	2 (28.5)	5 (71.4)	6 (75)	1 (50)	6 (46.1)	18 (69.2)	17 (50)
Major diseases										
Hypertension	5 (55.5)	3 (50)	5 (62.5)	4 (57.1)	3 (37.5)	6 (85.7)	—	9 (69.2)	13 (50)	22 (64)

(Contd.)

Table 4.2.3 : (Contd.)

Osteoporosis	—	1 (16.6)	2 (25)	1 (13.2)	—	2 (28.5)	—	5 (38.4)	2 (7.6)	9 (26.4)
Heart disease	1 (11.1)	2 (33.3)	3 (37.5)	1 (13,2)	—	1 (14.2)	2 (100)	4 (30.7)	6 (2.3)	8 (23.5)
Asthma	2 (22.2)	—	—	2 (28.5)	—	2 (28.5)	—	3 (23)	2 (7.6)	7 (20.5)
Diabetes mellitus	1 (11.1)	--	2 (25)	—	1 (12.5)	2 (28.5)	—	—	4 (15.3)	2 (5.8)
Cancer	—	—	1 (12.5)	—	—	—	—	—	1 (3.8)	—

Notes : Figures in parentheses represent percentage.
M =Male, F = Female

compared to females (14.2 %, 28.5%). In old age home IIIrd and IVth females had greater health problem than males.

Depression was major psychological health problem in both the sexes in all the old age homes. The elderly becomes ill more frequently, have to live with more chronic diseases or problems, and are generally trying to fight against several health problems at once, resulting in an increase in the number of medications taken.

4.2.3.2 Major disease

Major diseases were identified as hypertension, osteoporosis, heart disease, asthma, diabetes mellitus and cancer. It was found in old age home Ist, both the sexes had hypertension (55.5 %, and 50%), while 7.6 per cent, and 33.3 per cent females had osteoporosis, and heart disease. 22 per cent and 11.1 per cent males suffered from asthma, and diabetes mellitus, respectively. None reported severe diseases such as that of cancer. In old age home IInd, both the sexes had hypertension, osteoporosis, and heart disease, while only 25 per cent and 12.5 per cent males reported diabetes mellitus, and cancer. In old age home IIIrd, females had reported hypertension, osteoporosis, heart disease, asthma, and diabetes mellitus more prevalent than males. In old age home IVth, 100 per cent males had heart disease, while females (69.2, 38.4, 30.7, and 23%) had hypertension, osteoporosis, heart disease, asthma, and diabetes mellitus, respectively.

4.3 Age and sex-wise distribution of occurrence of accidents

Older persons, whose bones are often less dense and more brittle, are especially vulnerable to serious injuries from home accidents. For elderly people living independently without any assistance should consider all improving methods for the safety within their homes, as elderly people are more prone to serious injuries and long-term disability if they fall.

Home injuries occur as a result of falls, slips and most of these accidents are considered preventable. In poor or substandard housing, there may be an association between accidents and the greater number of hazards present. Table 4.3.1 revealed that the age and sex-wise distribution of total number of accidents in four old age homes. It was found that all accidents due to fall in the home (53.3%) which result to injuries occurred among elderly people belonged to the age group of 70-80 years, out of which 75 per cent were females. Females had a higher fall rate in all age groups and the severity of their injuries appeared to increase with age.

4.3.1 Place and type of occurrence of accidents in old age homes

Table 4.3.1 further illustrated the place, type of accidents due to fall, and also the causes of accidents in all the areas of the home. The study revealed that many accidental injuries occur in the bathrooms followed by separate toilets and corridors. Approximately 75 per cent of domestic accidents to people aged 75 and above were due to falls *(Wright, 1994)*. Furthermore, accidents may lead to a decline in self-confidence and increasing social isolation. Many accidental injuries occur in the bathrooms and can be avoided with common sense and inexpensive measures. We should identify bathroom hazards and find out ways to minimize them.

4.3.1.1 Causes of fall, slip and trip

In most of the cases the accidents were due to slips followed by falls. The causes of falls/slips were wet and slippery flooring. Tripping was not a cause of any single accidents in the studied group. Falls in the home are one of the most common accidents that can easily be prevented. A simple fall that results in a broken bone can become a sericus, disabling injury that limits ones independence. Smooth floor surfaces should be minimized, and the leaks should be corrected to avoid creating slippery floors, which is another cause of falls. Majority of falls occur in the home and surroundings, a review of hazards and modification to address those identified are an important component in falls prevention. The findings of chi-square calculated values showed that the occurrence of accidents depends on the age and sex of the elderly people.

Table 4.3.1 : Age and sex-wise distribution of occurrence of accidents

Old age home	60-70 years		70-80 years		80+ years		Total		Chi-square value
	M	F	M	F	M	F	M	F	
I[st]	—	1(16.6)	2(22.2)	—	—	1(16.6)	2(22)	2(33)	0.000NS
IInd	2(25)	—	—	1(14.2)	—	2(28.5)	2(25)	3(42)	1.701 NS
IIIrd *	—	2(25)	—	1(12.5)	—	1(12.5)	—	4(50)	4.120 *S
IVth*	—	—	—	3(23.0)	—	—	—	3(23)	3.931 *S
Total							**4**	**12**	

* Significant at 5 per cent levels of significance.
M = Male, F= Female
Notes : Figures in parentheses represent percentage.

Table 4.3.2 : Place and type of occurrence of accidents in old age homes (F=15)

Description	*Old Age Home*				*Total F = 60*
	*I^st^*	*IInd*	*IIIrd*	*IVth*	
I. Place					
a. Toilets (separate)	—	—	2 (13.3)	1 (6.6)	3 (5)
b. Bathroom with W.C.	3 (20)	2 (13.3)	2 (13.3)	1 (6.6)	8 (13.3)
c. Stairs	—	—	—	—	—
d. Corridors	—	2 (13.3)	—	—	2 (3.3)
e. Room	1 (6.6)	1 (6.6)	—	1 (6.6)	3 (5)
II. Type					
a. Fall	2 (13.3)	1 (6.6)	2 (13.3)	2 (13.3)	7 (11.6)
b. Slip	2 (13.3)	4 (26.6)	2 (13.3)	1 (6.6)	9 (15)
c. Trip	—	—	—	—	—
III. Causes of Fall					
a. Split flooring	—	—	—	—	—
b. Slippery flooring	1 (6.6)	1 (6.6)	2 (13.3)	—	4 (6.6)
c. Poor light	—	—	—	—	—
d. Poor health	1 (6.6)	—	—	2 (3.3)	3 (5)
IV. Causes of Slip					
a. Slippery flooring	1 (6.6)	1 (6.6)	1 (6.6)	—	3 (5)
b. Wet floor	1 (6.6)	1 (6.6)	—	1 (6.6)	3 (5)
c. Poor light	—	1 (6.6)	1 (6.6)	—	2 (3.3)
d. Poor health	—	1 (6.6)	—	—	1 (1.6)
V. Causes of Trip					
a. Split floor	—	—	—	—	—
b. Uneven surface	—	—	—	—	—

(Contd.)

Table 4.3.2 : (Contd.)

VI. Restriction to daily activities					
a. Limited	—	2 (13.3)	4 (26.6)	1 (6.6)	7 (11.6)
b. Fairly bit	6 (40)	5 (33.3)	2 (13.3)	7 (46.6)	20 (33)
c. Quite a lot	1 (6.6)	2 (13.3)	4 (26.6)	3 (20)	10 (16)
d. No restrictions	8 (53.3)	6 (40)	5 (33.3)	4 (26.6)	23 (26)

Note : Figures in parentheses represent percentage.

Moreover the female elderly people were more prone to accidental falls than males.

4.3.1.2 Restriction to daily activities

The physical abilities are reduced, and certain tasks such as stretching, lifting, and bending become more difficult for the elderly people. In old age home I^{st}, 53.3 per cent of the elderly people had no restrictions in their daily activities. In old age home IInd, 40 per cent had no restrictions in their daily activities, each 13.3 per cent had limited, and quite a lot restrictions in their daily activities. In old age home IIIrd, each 26.6 per cent had limited, and quite a lot restrictions in their daily activities. About 33.3 per cent had no restrictions but 13.3 per cent had fairly bit restriction to their daily activities. In old age home IVth, about 47 per cent elderly people had fairly bit restrictions followed by 26.6 per cent with no restrictions, 20 per cent with quite a lot, 6.6 per cent limited restrictions in their daily activities.

4.4 Dimension of old age homes

A home may generally be divided into number of discrete areas which include bedroom, kitchen, living room, bathroom and others. Each area has special functioning work task, and it is thus necessary to design functional areas which do not impede the profile of an ageing user.

4.4.1 Dimension of rooms in old age homes

4.4.1.1 Dimension of habitat room

The old age homes contain a living room (*i.e.* one habitat room used for living, sleeping, or eating purposes by single elderly people). The findings in Table 4.4.1 showed that the length, width and height of the habitat room of old age home Ist were 5.1 × 3.0 × 3.9 metres, respectively. Corresponding dimensions of old age home IInd, IIIrd and IVth were 3.0 × 3.3 × 3.3 metres, 5.4 × 3.2 × 3.2 metres and 3.2 × 3.2 × 3.5 metres, respectively. The rooms of Ist, and IIIrd old age homes were bigger than that of IInd. In old age home IVth there was a dormitory with size 6.3 × 5.3 metres also for accommodating six elderly people. The average length, width, and height of habitat rooms in all the old age homes were 4.17 × 3.17 × 3.47 metres; with standard deviation 1.25, 0.12, and 0.30, respectively. As per recommendation given in National Building Codes, the room size should not be less than 9.5 m² with a minimum width of 2.4 metres. The rooms of all the old age homes were within the recommended value.

4.4.1.2 Bath with water closet

A working toilet must be available for the exclusive use of the occupants of the unit. In old age home Ist, IInd, IIIrd and IVth the length, width, and height of bathrooms with water closets were 1.3 × 1.8 × 3.9 metres, 1.5 × 1.8 × 3.3 metres, 1.8 × 1.5 × 3.2 metres, 1.5 × 1.0 × 3.5 metres, respectively. The mean length, width, and height of bath and water closet were 1.52 × 1.52 × 3.4 metres with standard deviation 0.20, 0.37, 0.30, respectively. As per the recommendation given in National Building Codes, the bath with water closet, the floor area shall not be less than 2.8 m² with a minimum width of 1.2 metres. The baths with water closets of all the old age homes were within the recommended value.

4.4.1.3 Separate water closet

Water Closet (or W.C), is a small room or booth containing a toilet,

Table 4.4.1 : Dimensions of old age homes (metres)

Description	*Old age home*				*Mean*	*S.D.*	*Fisher's t-value*
	I	*II*	*III*	*IV*			
Habitat room							
Length	5.1	3.0	5.4	3.2	4.17	1.25	1.2825
Width	3.0	3.3	3.2	3.2	3.17	0.12	
Height	3.9	3.3	3.2	3.5	3.47	0.30	
Bath with W.C							
Length	1.3	1.5	1.8	1.5	1.52	0.20	1.767
Width	1.8	1.8	1.5	1.0	1.52	0.37	
Height	3.9	3.3	3.2	3.5	3.4	0.30	
Separate W.C*							
Length	—	—	—	1.6	1.6	—	21.896 *
Width	—	—	—	1.5	1.5	—	
Height	—	—	—	3.5	3.5	—	
Separate Bath							
Length	—	—	—	1.6	1.6	—	—
Width	—	—	—	1.5	1.5	—	—
Height	—	—	—	3.5	3.5	—	—
Kitchen							
Length	5.4	6.0	5.5	5.1	5.5	0.37	—
Width	3.6	3.8	3.2	3.2	3.45	0.3	—
Height	3.9	3.3	3.2	3.5	3.47	0.30	—
Dinning Spaces							
Length	7.5	3.0	5.7	6.0	5.55	1.87	—
Width	7.5	2.6	3.3	2.1	3.87	2.46	—
Height	3.9	3.3	3.2	3.5	3.47	0.30	—
Store room*							
Length	—	—	—	4.1	4.1	—	176.48 *
Width	—	—	—	2.3	2.3	—	—
Height	—	—	—	3.5	3.5	—	—

* Significant

and often a sink in the unit that provides for privacy. In old age home IVth, there was provision of separate water closet with size 1.6 × 1.5 × 3.5 metres

4.4.1.4 Separate bathroom

At least one working bathroom is required for the exclusive use of the occupant. It should be a soothing and comforting place, offering health, relaxation and cleanliness. In old age home IVth, there was provision of separate bathrooms with size 1.6 × 1.5 × 3.5 metres.

4.4.1.5 Kitchen

All units are required to contain a kitchen for the storage and preparation of food. Space for preparation of food (counter tops, etc.) must be free of damage, holes, and lifting surfaces that allow contamination, food build up, or insects. The size of the common kitchen in old age home Ist was 5.4 × 3.6 × 3.9 metres in length, width and height, respectively. Corresponding dimensions of old age home IInd, IIIrd and IVth were 6.0 × 3.8 × 3.3 metres, 5.1 × 3.2 × 3.2 metres, and 5.5 × 3.2 × 3.5 metres, respectively.

4.4.1.6 Dinning spaces

Dinning spaces should not disturb the activities in living, and other private areas. In old age home Ist there was a well maintained of dinning space which was of 7.5 × 7.5 × 3.9 metres. There was no separate dinning space in old age home IInd, but the veranda was used as dinning purposes. The length, width, and height of this were 3.0 × 2.6 × 3.3 metres, respectively. Corresponding dimensions dinning spaces of old age home IIIrd and IVth were 5.7 × 3.3 × 3.2 metres and 6.0 × 2.1 × 3.5 metres, respectively. The average length, width, and height were 5.5 × 3.87 × 3.47 metres, respectively.

4.4.1.7 Store room

There was separate store room for keeping personal belonging those who are living in dormitory in old age home IVth, and the size of store room was 4.1 × 2.3 × 3.5 metres. The findings of Fisher's *t*-values showed that the dimensions of habitat rooms were found in accordance with National Building Codes, whereas the separate water closets and store rooms were different from National Building Codes.

4.4.2 Dimension of storage spaces in old age homes

All the old age homes are provided with storage spaces for keeping their personal belonging. In general, the requirements of elderly people for clear visibility and easy reach should determine the placement of storage shelves in the home. A storage option is storage above shoulder height (for light, less frequently used items) and the designing criteria is shoulder height from floor in standing position, accessible by 5th percentile female, *i.e.* 117.75 cm. The present study (Table 4.4.2) showed the 5th percentile female shoulder height was 117.75 cm and the average maximum height from floor of cupboard was 186 cm. It showed that the majority of elderly people had problems in reaching the top shelf of the cupboard of the higher facilities. None of the elderly people were able to straighten their ankles totally any more. A storage option for shelf height for items requiring visual inspection by the user and the designing criteria is standing eye height from floor, 5th percentile female *i.e.* 128.87 cm. Though the mean upward reach of the elderly males and females was 217.5 cm, 173.08 cm, respectively, for the right hand, most of the elderly females complained of the height of the top shelves.

The results showed that the minimum height from floor of built in cupboards and open racks was 31 cm and 22 cm, respectively. Storage option for lower shelves (medium to heavy weight items) and the design criteria should be standing knuckle height from floor, 95th percentile male. So that elderly users are not subjected to bending and overreaching demands. The decrease in functional abilities of elderly individuals prevents them from stretching to reach surfaces. 50 per cent of the elderly people also had problems with the lower shelves of cabinet with two shelves (10 cm). The above-work surface cupboards may be too high to be of much use and the below-work surface cupboards too low. *Grandjean (1973)* recommended that the kitchen shelves should have as much adjustment as possible within the range of 300-1600 mm

Table 4.4.2 : Dimensions of storage spaces in old age homes (cm)

Description	*Maximum Height*	*Height from floor to lower shelves*	*Shelves Depth*	*Shelves Length*	*Material used*
1. Built in cupboard					
Old age home I^st	180	23	42	82	Wooden
Old age home IInd	200	54	45	120	Wooden
Old age home IIIrd	175	28	43	88	Wooden
Old age home IVth	190	21	50	190	Wooden
Mean	186	31	45	120	
S.D.	0.44	0.15	0.03	0.49	
2. Open rack					
Old age home I^st	—	—	—	—	—
Old age home IInd	145	23	28	71	Concrete
Old age home IIIrd	—	—	—	—	—
Old age home IVth	121	21	28	87	Concrete
Mean	133	22	28	79	
S.D.	0.16	0.01	0	0.11	
3. Cabinet with two shelves					
Old age home I^st	—	—	—	—	—
Old age home IInd	—	—	—	—	—
Old age home IIIrd	73	10	50	50	Wooden
Old age home IVth	80	11	30	30	Wooden
Mean	76	10	40	40	
S.D.	0.04	0.007	0.14	0.14	

above the floor. *Pheasant (1988)* gives the following figures for predicted optimal heights of base unit surfaces for adults: men (50th percentile) 1015 mm and females (50th percentile) 930 mm. *Kothiyal,K. and Tettey, S. (2001)* recommended values for some storage options at work and home environment as under:

Storage option	*Design criteria*	*Recommended design values (cm)*
Storage above shoulder height (for light, less frequently used items).	Shonlder height from floor, standing, accessible by 5th percentile female with a 20° (maximum) joint flexion.	135 cm (maximum)
Shelf height for items requiring visual inspection by the user.	Standing eye height from floor, 95th percentile female.	133 cm
Lower shelves (medium to heavy weight items).	Standing knuckle height from floor, 95th percentile male.	63-65 cm

The findings of chi-square calculated values showed no significant relationship between the anthropometric data of elderly people and the dimensions of storage spaces. It is inferred that the dimensions of storage spaces were independent of anthropometric measurements. In other words, the storage spaces were not made according to the dimensions of human body.

4.4.3 Dimensions of door, window and ventilator

A well designed door, window and ventilator should let in ample sunlight, and open for. natural ventilation. It also gives excellent sealing and complete insulation from sound, heat, and dust.

4.4.3.1 Door in habitat room

Doors may be equipped with a night latch, hook and eye, slide bolt or security chain provided such devices are openable from the inside without the use of a key or tool, and mounted at a reasonable height for the occupant. Finding, summarized in Table 4.4.3 showed that the length and width of doors in habitat rooms were 1.8 × 1.0 m, 1.8 × 0.9 m, 2.0 × 0.72 m, and 0.97 × 0.56 metres in Ist, IInd, IIIrd, and IVth old age homes, respectively. The mean length and width of doors in habitat rooms was 1.64 and 0.79 metre with 0.45 and 0.19 S.D. Door shall have a clear opening of not less than 80 cm when open, and shall be operable by a single effort. It was found to be within this recommendation.

4.4.3.2 Window in habitat room

All bedrooms or any other rooms used for sleeping must have at least one openable window. Acceptable under certain conditions are windows in rooms used for sleeping that open onto common areas rather than to the outside. These are acceptable facilities provided through the windows that allow adequate natural light to permit normal indoor activities. The length and width of windows were 1.2 × 1.2 m, 1.5 × 0.9 m, 1.0 × 0.8 m, 1.1 × 0.9 metres in

Table 4.4.3 : Dimensions of door, window and ventilator (metres)

Descriptions	*Old age home*				*Mean*	*S.D.*
	I	*II*	*III*	*IV*		
Habitat room						
Door						
Length	1.8	1.8	2.0	0.97	1.64	0.45
Width	1.0	0.9	0.72	0.56	0.79	0.19
Material used	Wooden	Wooden	Wooden	Wooden		
Window						
Length	1.2	1.5	1.0	1.1	1.2	0.21
Width	1.2	0.9	0.8	0.9	0.9	0.08
Material used	Wooden	Wooden	Wooden	Wooden		
Ventilator						
Length	0.46	0.46	0.47	0.4	0.44	3.20
Width	1.00	0.61	1.0	1.1	0.92	0.22
Material used	Wooden	Wooden	Wooden	Wooden		
Bath with W.C.						
Door						
Length	1.8	1.8	1.95	0.97	1.63	0.38
Width	0.63	0.70	0.65	0.56	0.63	0.05
Material used	Wooden	Wooden	Wooden	Wooden		
Ventilator						
Length	0.46	0.46	0.47	0.4	0.44	3.20
Width	1.00	0.61	1.0	1.1	0.92	0.22
Material used	Wooden	Wooden	Wooden	Wooden		

Ist, IInd, IIIrd, and IVth old age homes, respectively. The mean length, and width of windows in habitat rooms was 1. 2 × 0. 9 metres with 0.21 and 0.08 S.D. It showed that one window was large enough to provide natural lighting and air circulations in the habitat room.

4.4.3.3 *Ventilator in habitat room*

The exchange between indoor and outdoor air profoundly affects contaminant levels in a dwelling *(Esmen, 1985)*. The concentration of pollutants depends upon rates of production and removal, the source, and their dilution by ventilation *(Lowry, 1989)*. Adequate ventilation is imperative to discourage humidity from reaching unacceptable levels *(Collins, 1993)*. It was found that the length and width of ventilators were 0.46 × 1.0 m, 0.46 × 0.61 m, 0.47 × 1.0 m, and 0.40 × 1.1 metres in Ist, IInd, IIIrd and IVth old age homes, respectively. The mean length and width of ventilators was 0.44 × 0.92 metres with 3.25 and 0.22 S.D. The ventilator or window shall have an open space of area not less than 0.3 m^2 with side not less than 0.3 metres. It was found to be within this recommendation.

4.4.3.4 Door in bath with water closet

The doors of the bath with water closets were smaller in width than habitat rooms. The length and width of doors of bath with water closets were 1.8 × 0.63 m, 1.8 × 0.70 m, 1.9 × 0.65 m, and 0.97 × 0.56 metres in Ist, IInd, IIIrd and IVth old age homes, respectively. The mean length and width of door was 1.63 × 0.63 metres. The average width of the doors of bath with water closets was 0.63 cm which was observed to be far less than the standard recommended width of 80 cm. Hence, it may be caused inconvenience to the elderly people.

4.4.3.5 Ventilator in bath with water closet

In order to prevent the accumulation of unhealthful odors, sewer gas, and the build up of mould on surfaces, an openable window or a working ventilation system is required in bathrooms. Healthy indoor environments with adequate fresh air ventilation were also proposed as a way to manage mite-sensitive asthma *(Harving et al., 1994)*. If a window is to be used for bathroom ventilation, a screen is required. The size of the ventilators in all the old age homes was identical as in the habitat rooms. The types of ventilation systems that are allowable are electric vent fan either wall or ceiling mounted, gravity flow vents, and shafts that allow air to escape to the outside.

4.5 Description of infrastructure facilities available in old age homes

The infrastructural facilities in the old age homes were quantified by assigning scores for Yes statements 1 and for No statements 0, with the total score, the percentage was computed. The scores were interpreted such that higher the score the greater the infrastructural facilities available.

4.5.1 Furniture and furnishing

Furniture includes chair, table, bed etc. it should be soothing to eyes. Over-crowding of furniture should be avoided to enhance positive energy inside. Furnishings should be appealing, functional, and comfortable. Table 4.5 revealed that in old age home Ist, the furniture and furnishing provided were not sufficient as per the requirement of elderly people. There was no special relaxing chair where the elderly people are supposed to sit for a long time. Regarding furnishing nothing was being provided, the elderly people had to arrange everything themselves. A well maintained dinning room was available but it was seldom used by them. In old age home IInd, the furniture and furnishing was totally provided by the authority but it was not up to the requisite requirement of the elderly people. In old age home IIIrd, the quantity of furniture available to them was superior to other old age homes. Little furniture like easy chair was made available. The furnishing like curtains, bed sheets, pillow covers were also taken care of and were changed frequently. Here also a well maintained dinning room was also available but it was seldom used. In old age IVth, the furniture and the furnishing were not enough, and up to the requirement for the inmates.

4.5.2 Fittings and fixtures

A building may have following fixtures and fittings like door bell, knocker, handles, bolts, wall fittings (shelves/mirrors etc.) curtains, carpet, cupboards, electricity fittings—bulbs, fluorescent tubes,

Table 4.5 : Description of infrastructure facilities available in old age homes

Description	*Old Age Home*			
	I *Score*	*II* *Score*	*III* *Score*	*IV* *Score*
I. Furniture and furnishing				
a. Sufficient number of furniture	1	1	1	1
b. Special kind of easy chair for relaxing	0	0	1	0
c. Mattress and cushion	1	1	1	1
d. Dinning table facility	1	0	0	0
e. Rubber mates placed in areas likely to get wet.	0	0	0	0
II. Fittings and fixtures				
a. Handholds and grab bars in sleeping, bathroom, toilet	0	0	1	0
b. Lighting in stairs, bathroom, corridor	0	0	0	0
c. Larger switches which glow-in-light	0	0	0	0
d. Coloured signal lights which attract attention to a hazard	0	0	0	0
e. Wash-basins	1	0	1	1
f. Special toilet seats with arm rest or side bar	0	0	1	1
III. Medical facilities				
a. Emergency bell	0	0	1	0
b. Regular health check-up	0	0	0	0
c. First aid facility for urgency	1	0	1	1
d. Regular visits of Doctors	0	0	0	0
e. Ambulance facility for an emergency	1	1	1	1
f. Free medicine	0	1	0	0

(Contd.)

Table 4.5 : (Contd.)

IV. Any other facilities				
a. Recreational facilities —T.V. Radio, Library, Temple	1	1	1	1
b. Generator	1	1	0	0
c. Phone facilities in room	0	0	0	0
Total score : 20	8 (40)	6(30)	10(50)	7(35)

Yes = 1
No = 0

switches, bathroom and water closets fittings—taps, exhaust fan, basin, and handholds or grab bars. In old age home Ist the fittings and fixtures such as handholds, bars, larger electric switches, convenient wash basins, taps etc. were lacking. In old age home IInd had similar plight of fittings and fixtures as above old age home, but the few rooms where elderly persons who were in lacking capability to physical strength had provided handholds in bath with combined water closet. In old age home IIIrd, the fittings and fixtures were satisfactory to some extent like handholds in bathroom and electric switches close to bed in reachable height but the sufficient lighting fixtures were found missing to provide adequate light at night hours. In old age home IVth, most of the fittings and fixtures such as handholds, bars, larger electric switches, convenient wash basins, taps etc. were lacking. The inmates who were living in dormitory have to use the common bath and water closet. The availability of a toilet on the same level as the living room might be a luxury for younger ones but it assumes greater significance for elderly people to have a bathroom or at least a toilet in the living area. The electric vent fan ventilators were not found in all the old age homes however, there was a working vent systems or operable ventilator.

4.5.3 Medical facilities

A medical facility includes regular visits to old age homes by geriatric doctors and nursing care such as dressing, medication and injection etc. It was found that old age home Ist, and IInd had their hospitals within the proximal distance. The elderly people had to bear the cost of treatment and medicines themselves in all the old age homes except in old age home IInd. However, the poorest inmates were taken care of by the authorities. In old age home IIIrd and IVth no ambulance facilities was available at the time of emergency. There was no personal emergency response system like bell for immediate assistance in the event of an emergency in any of the old age homes except in old age home IIIrd

4.5.4 Any other facilities

Social, recreational and educational activities provided to develop their potential to promote a family atmosphere and to enhance mutual support among elderly people. It was found that the recreational facilities like television, radio were not provided to the inmates and they themselves had to arrange for it, though news-paper, were made available in all the old age homes. The old age home I[st] and IInd were within the vicinity of temple and those who were in good physical strength had joined the schedule activities of morning yoga while other, involved themselves in religious activities like *puja, aarti prawachan* and so forth. The elderly people also found engaged themselves in activities like the male graduates took the charges of account keeping and other official work and the females were helping in meal preparation like cutting vegetables, rolling chapattis etc. The library and church were located within the vicinity of the old age home IIIrd and IVth and the inmates joined there to keep themselves engaged. There was a generator facility at the time of light cut off in the old age home I[st] and IInd but it was restricted during day hours, however, it was supplied to some important areas like corridors. Not a single old age homes had provided phone in rooms, however, there was phone in the office. The scores were highest 50 per cent in old age home IIIrd which showed that the infrastructural facilities was superior followed by old age home I[st], 40 per cent, and in old age home IVth 35 per cent, while lowest 30 per cent facilities were available in old age home IInd as it was free of cost.

4.6 Dimensions of furniture

It includes the length, width, depth of the chair, table, and bed provided in the old age homes.

4.6.1 Dimension of chair

Chair design considers intended usage, ergonomics (how comfortable is it for the occupant), as well as non-ergonomic functional requirements such as size, stack ability, fold ability, weight, durability, stain resistance and artistic design. In general, if the occupant is suppose to sit for a long time, weight needs to be

Table 4.6.1 : Dimensions of chair (cm)

Old age home	*Total Height*	*Seat (from floor)*			*Back rest*		*Angle of back rest*	*Material used*
		*Height**	*Width*	*Depth**	*Height*	*Width*		
I[st]	84	43	42.5	42.5	46	47	105°	Plastic
IInd	85	47.2	47.2	35	43	46	100°	Plastic
IIIrd								
Chair first	64	46	47.5	50	53	45	90°	Wooden
Chair second	85	45	50	39	46	43	114°	Wooden
IVth	84	44	45.5	42	45	45.5	110°	Plastic
Mean	80.4	45.04	46.54	41.7	46.6	45.3	103.8°	
S.D.	9.18	1.64	2.77	5.51	3.78	1.48	9.33°	
Fisher's *t*-value	—	12.283* S	—	5.794*S	—	—	—	—

* Significant

taken off the seat area and thus "easy" chairs intended for long periods of sitting are generally at least slightly reclined. Nearly identical chairs were found in old age home I[st] and old age home IInd. The average total height, seat height from floor, width and depth of the seat of the chair were 80.4 cm, 45.04 cm, 46.54 cm, and 41.7cm, respectively. The mean height and width of back rest was 46.6 cm and 45.3 cm, respectively. The angle of the back rest of wooden chair was 90°. The mean angle of the back rest was 103.8°. Chair with cushioning and backward slop of 115° help in decreasing postural discomfort, and increase efficiency. A foot rest must be attached to the chair so that the elderly person can place feet flat to keep knee joint angle at approximately 90°. Chairs should also be of an appropriate height to allow the individual's feet to rest flat on the floor with no pressure behind the knees. The recommended *(IS: 5974-1986 Reaffirmed 1996)* dimensions of easy chairs are as follows:

Seat height	38-40 cm
Seat depth	52-60 cm
Height of armrest above seat	15-23 cm

The chairs in present study were not found within the recommendation.

4.6.2 Dimension of bed

The length, width and height from floor of bed of old age home I[st], IInd, IIIrd and IVth were 190 × 94 × 46.0 cm, 177.5 × 87.5 × 46.25 cm, 201 × 92 × 46 cm, and 193 × 90 × 46.2 cm, respectively. The average length, width and height from floor of bed were 190 cm, 37 cm, 90.87 cm, and 46.11 cm, respectively with S.D. ranges from 0.13 to 9.75. The recommended [IS: 7259 (Part 1)–1988] overall dimensions of the single bed shall be as follows:

(a)	Height from floor to the top of the mattress	Maximum 45 cm
(b)	Clear in-bed length between head boards and the foot boards	190-200 cm
(c)	Width	91 cm

Table 4.6.2 : Dimensions of bed (cm)

Old age home	*Length**	*Width*	*Height from floor*	*Material used*
Ist	190	94	46.0	Wooden
IInd	177.5	87.5	46.25	Wooden
IIIrd	201	92	46	Iron
IVth	193	90	46.2	Iron
Mean	190.37	90.87	46.11	
S.D.	9.75	2.78	0.13	
Fisher's *t*-value	5.200*S	0.899** NS	—	

* Significant
**Not significant

The measurements of bed provided in old age homes were not found within this recommendation. The computed Fisher's *t*-value showed the significant relationship for height, depth of chair, whereas for width of bed not significant. It is inferred that the chair and bed was not constructed according to the recommendations given for furniture codes of Indian standard.

4.6.3 Dimension of table

Tables, bookcases, and other furniture should be sturdy and able to withstand weight from persons leaning for support. The length, depth and height of knee zone of the table used by the elderly people in Ist, and IVth old age homes were 42.5 × 80 × 55 cm and 122 × 61 × 71 cm, respectively. In old age home IInd and IIIrd the length, depth and height of knee zone of the table first were 180 × 60 × 70 cm and 157 × 40 × 71 cm, respectively and table second were 180 × 35 × 25 cm and 122 × 74 × 76 cm, respectively. The average length, depth, and height of the knee zone were 133.91 cm, 58.33 cm, and 61.33 cm, respectively with S.D. ranging from 17.91 to 51.80. In case of table height it should correspond to seated elbow height or 95th percentile of elbow rest height. Data regarding elbow height of selected elderly people in further section (Tables 4.7.1&2) showed that the 95th percentile of elbow rest height from seat was 23.75 cm. and adding ideal seat height to it gives 66.6 cm of table height, while existing table height in all four old

Table 4.6.3 : Dimensions of table (cm)

Description		Length	Depth	Height of knee zone	Material used
I[st]	Table Small	42.5	80	55	Wooden
IInd	Table first	180	60	70	Wooden
	Table second	180	35	25	Wooden
IIIrd	Table first	157	40	71	Wooden
	Table second	122	74	76	Wooden
IVth		122	61	71	Wooden
Mean		133.91	58.33	61.33	
S.D.		51.80	17.91	19.16	

age homes were more than this value. It showed that the existing tables did not match with most of the elderly people. A higher than usual seat would impact on the clearance between the thigh and the underside of the table required for the free movement of legs. Therefore, table height for the elderly people would need to be accordingly adjusted. The height of the work table depends on the nature of the activity. For example, observations on the general working population suggest that writing is better done at a table height a little higher than elbow height. Work involving a moderate amount of force requires the working height to be 50 to 100 mm below the elbow *(Grandjean, 1988).* There is lack of data on preferred working heights for the elderly population, however, observations on the young population can be considered as a guide. The findings of chi-square calculated values showed no significant relationship between the anthropometric data of elderly people and the dimensions of chair and bed provided in old age homes. It is inferred that the dimensions of chair and bed were independent of anthropometric measurements. In other words, the furniture was not made according to the dimensions of human body.

4.7 Anthropometric measurements of elderly people

Anthropometric measurements of elderly people were taken in both sitting and standing positions [Fig. 4.4(a) and 4.4(b)]. Measurements were made with elderly people wearing light clothing and with bare feet. All elderly people were of normal physical health and were active in life at the time of the measurements. An adequate description of the human body may require over 300 measurements

Table 4.7.1&2 : Sex-wise percentile distribution of anthropometric dimensions of elderly people

Anthropometric measurements (cm) *STANDING***	*Male*					*Female*				
	5th	*50th*	*95th*	*S. D.*	*Average*	*5th*	*50th*	*95th*	*S. D.*	*Average*
Height (cm.)*				2.35	154.16				15.08	141.60
Weight (kg.)*				13.96	68.3				6.42	52.3
Body surface area (m^2)*				0.09	1.16				0.07	0.96
Eye height from ground (cm)	152.4	157.75	159.75	3.32	156.55	128.87	148	152.95	10.56	144.51
Shoulder height from ground (cm)	141	144.95	146.77	2.54	144.6	117.75	130.7	136.5	8.15	128.41
Elbow height from ground (cm)	107.15	110	116.27	3.83	110.81	95	100	102.62	3.15	99.41
Knee height from ground (cm)	52.45	57.65	58.85	2.93	56.75	44.37	50	52.97	3.58	49.43
Poplliteal height above ground (cm)	42.85	46.4	50.25	3.01	46.43	40.7	42.4	44.12	1.41	42.31
Upward reach (cm)	212.25	218.5	220.75	3.61	217.5	170.5	172.6	175.8	2.20	173.08
Sideways reach (cm)	69.07	71.8	73.5	1.88	71.58	64.5	68.75	70.75	2.61	68.08
Forward reach (cm)	72.8	74.7	76.35	1.49	74.73	61.25	63.25	70.4	4.03	64.45
Shoulder –elbow height (cm)	32.2	33.55	35.87	1.58	33.9	29.3	30.6	31.37	0.90	30.38
Forearm-elbow height (cm)	42.125	42.95	43.42	0.55	42.85	38.02	38.85	39.45	0.61	38.76
Hand length (cm)	19.125	20.25	22.47	1.39	20.55	15.65	17.45	18.75	1.26	17.35
Elbow to Elbow breadth (cm)	47.47	49.75	50.87	1.42	49.41	41	43.5	47.65	2.98	43.86
Total arm length (cm)	69	72.2	73.67	2.02	71.78	56.5	63.5	65.75	4.04	62
SITTING** (cm)										
Body height from ground (cm)	118.52	120.75	122	1.90	120.5	114.5	120.75	121.95	3.37	119.21
Body height above seat (cm)	73.82	75.65	76.9	5.91	75.5	72.52	73.35	73.87	0.55	73.25
Eye height from ground (cm)	108.3	110.87	110.97	1.90	110.08	106.7	109.75	110.37	1.72	108.9
Eye height above seat (cm)	64.625	66.45	68.12	1.43	66.41	61.17	62.5	64.45	1.36	62.7
Shoulder height from ground (cm)	90.5	99.35	101.25	4.70	97.45	89.85	96.2	98	3.57	94.63
Shoulder height above seat (cm)	54.125	61.7	65.75	5.14	60.65	49.42	50.8	51.92	1.05	50.73
Elbow height from ground (cm)	62.875	68.95	70.75	3.58	68.06	65.17	66.95	69.32	1.69	67.08

(Contd.)

Table 4.7.1&2 : (Contd.)

Elbow height above seat (cm)	21.25	22.75	23.75	1.02	22.58	17.025	18.05	19.17	1.08	18.08
Knee height above ground (cm)	51.25	53.1	53.87	1.11	52.78	44.12	46.9	48.72	1.93	46.8
Popliteal height above ground (cm)	40.35	43.2	43.95	1.65	42.76	35.75	37.75	39.25	1.44	37.61
Buttock popliteal lengths (cm)	48.525	50.75	51.37	1.23	50.25	46.07	48	49.65	1.50	47.98
Buttock knee lengths (cm)	55.525	56	56.37	0.35	55.96	49.22	52.5	53.62	1.95	52.06
Thigh heights (cm)	11.8	14.3	15.15	1.46	13.83	14.12	15.1	15.52	0.58	14.93

*Significant at 5% level of significance,
**Significant at 1% level of significance.

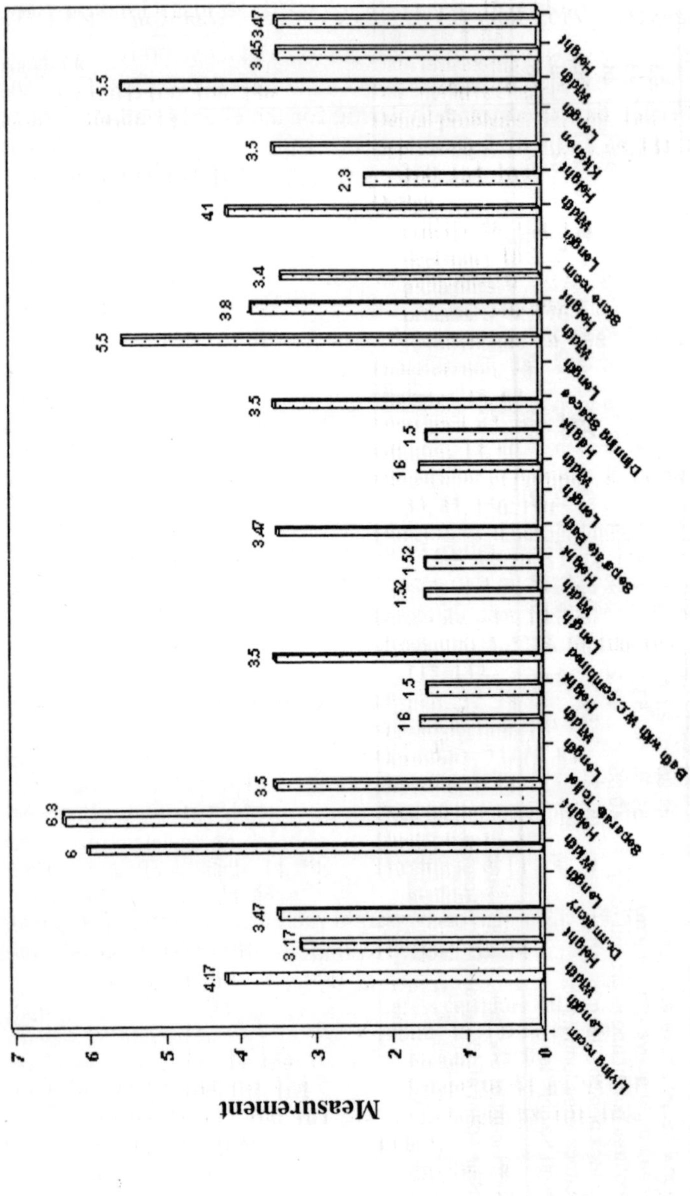

Fig. 4.2 : Dimensions of old age homes

(Pheasant 1986; Roebuck, Kroemer, and Thompson, 1975). The scope of this study was limited to measuring those body dimensions that were considered important and useful for the design of facilities and equipment used for daily living for the elderly and also for the design of the workplace environment *(Steenbekkers and Beijsterveldt, 1998).* In all 30 body dimensions were selected for measurements. The height, weight, body surface area, percentile of 5th, 50th, 95th as well as standard deviation, average values are presented in Table 4.7.1. Anthropometric data must be taken into consideration when the adequacy of furniture and fixtures is assessed.

4.7.1 Anthropometric measurement (standing position)

4.7.1.1 Height

Designers often use stature (body height) as an important reference value and try to relate it to other body dimensions to get approximate values for the purpose of designing items. The findings revealed that the elderly males had significantly higher height than the elderly females. The mean height of males and females was 154.16 cm and 141.6 cm with differences varying from 2.3 to 15.8 S.D. [Fig. 4.3(b)]. Studies have shown that the height of every adult individual declines with age *(Borkan et al. 1983; Damon and Stoudt 1981).*

4.7.1.2 Weight

As the age increases the body weight increases to certain proportion than becomes constant. Body weight of the elderly people was measured with standard weighing balance with minimum clothes and without shoes. The findings revealed that the elderly males had significantly higher weight than the elderly females. The mean weight of males and females was 68.3 kg and 52.3 kg, respectively with difference varying from to 6.4 to 13.96 S.D. [Fig. 4.3(a)].

4.7.1.3 Body surface area

The mean body surface areas of elderly males were higher than

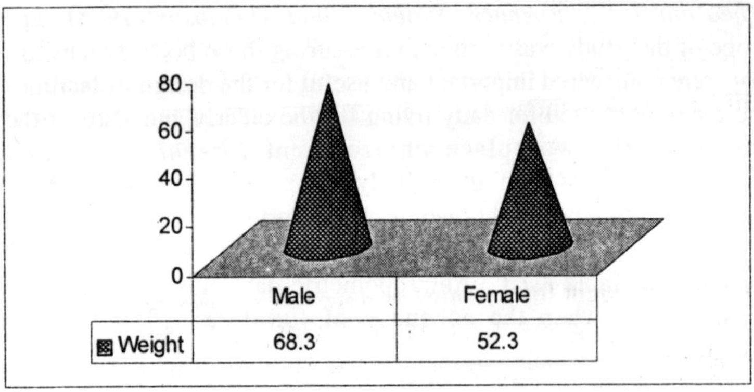

Fig. 4.3(a) : Sex-wise weight of elderly people

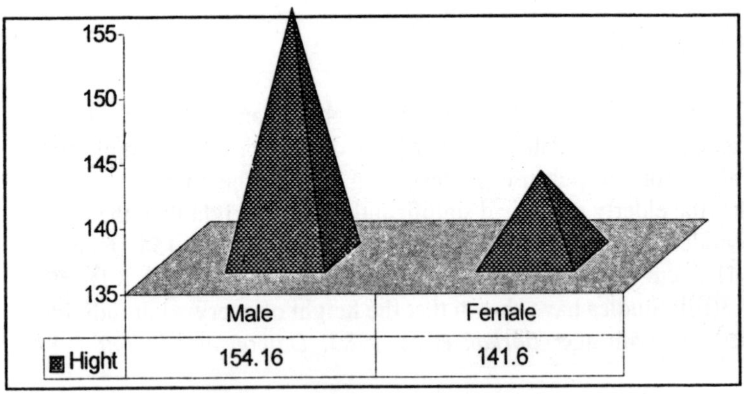

Fig. 4.3(b) : Sex-wise height of elderly people

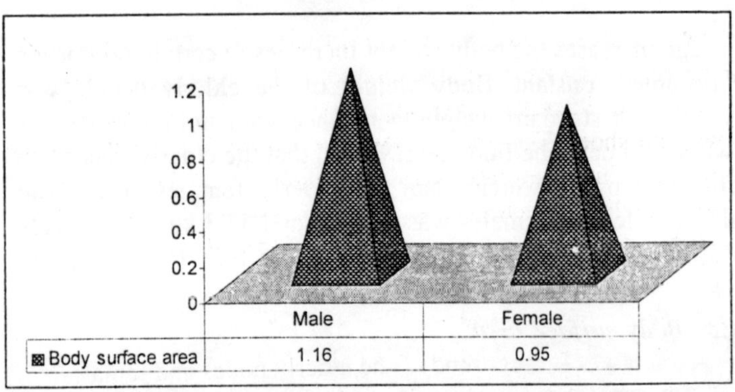

Fig. 4.3(c) : Sex-wise body surface area of elderly people

that of elderly females. The average body surface area of males and females was 1.16 and 0.95 square metres, respectively (Fig. 4.3(c)]. The findings of correlation co-efficient (*r* value) showed significant relations between the height, weight and body surface area of male and female elderly people at 5% levels of significance.

4.7.1.4 Eye height from ground

The mean eye height from ground of the elderly males and females was 156.55 cm, and 144.51 cm, respectively. The 5th, 50th and 95th percentile of elderly males and females were found to be 152.4 cm, 157.75 cm, 159.75 cm, respectively and 128.87 cm, 148 cm, 152.95 cm, respectively with difference varying from 3.32 to 10.56 S.D.

4.7.1.5 Shoulder height from ground

The mean shoulder height from ground of the elderly males and females was 144.6 cm and 128.41 cm, respectively. The 5th, 50th and 95th percentile of elderly males and females were found to be 141 cm, 144.95 cm, 146.77 cm, respectively and 117.75 cm, 130.75 cm, 136.5 cm, respectively with difference varying between 2.54 to 8.15 S.D.

4.7.1.6 Elbow height from ground

The mean elbow height from ground of the elderly males and females was 110.81 cm and 99.41 cm, respectively. The 5th, 50th and 95th percentile of elderly males and females were found to be 107.15 cm, 110 cm, 116.27 cm, respectively, and 95 cm, 100 cm, 102.62 cm, respectively with difference varying from 3.83 to 3.15 S.D. Elbow height has been said to be an important reference value for the determination of work surface height *(Pheasant, 1988).*

4.7.1.7 Knee height from ground

The mean knee height from ground of the elderly males and females was 56.75 cm and 49.43 cm, respectively. The 5th, 50th and 95th

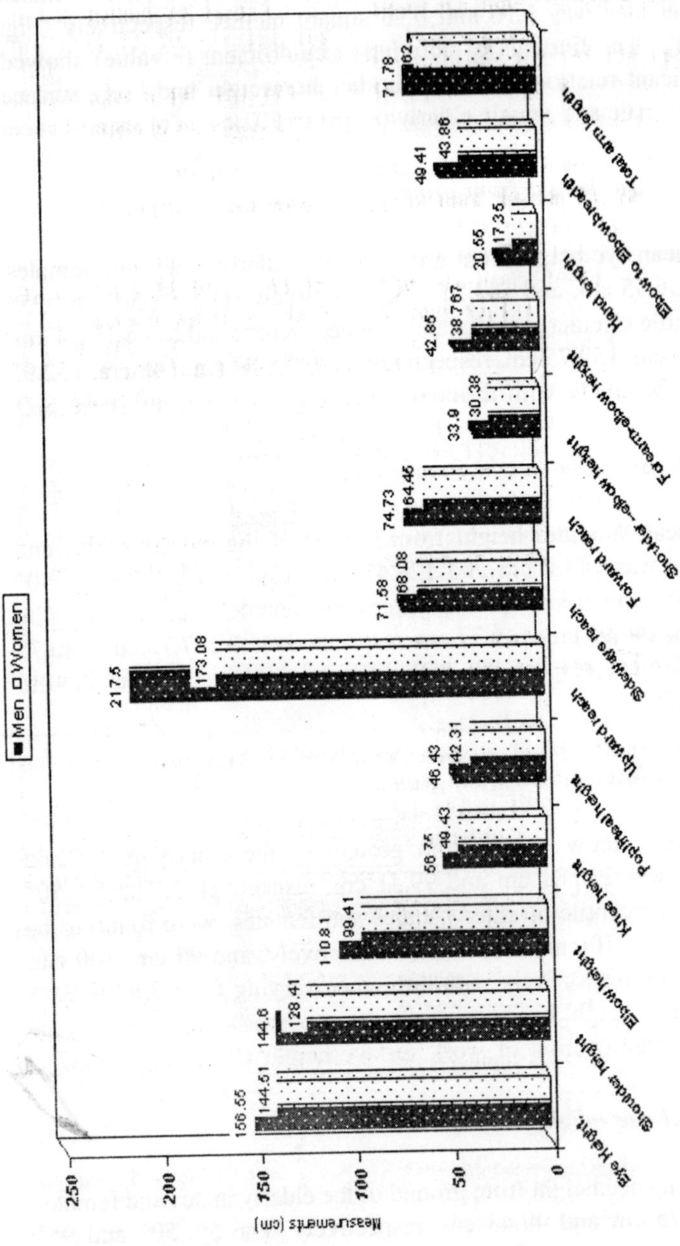

Fig. 4.4(a) : Sex-wise anthropometric measurements (standing position)

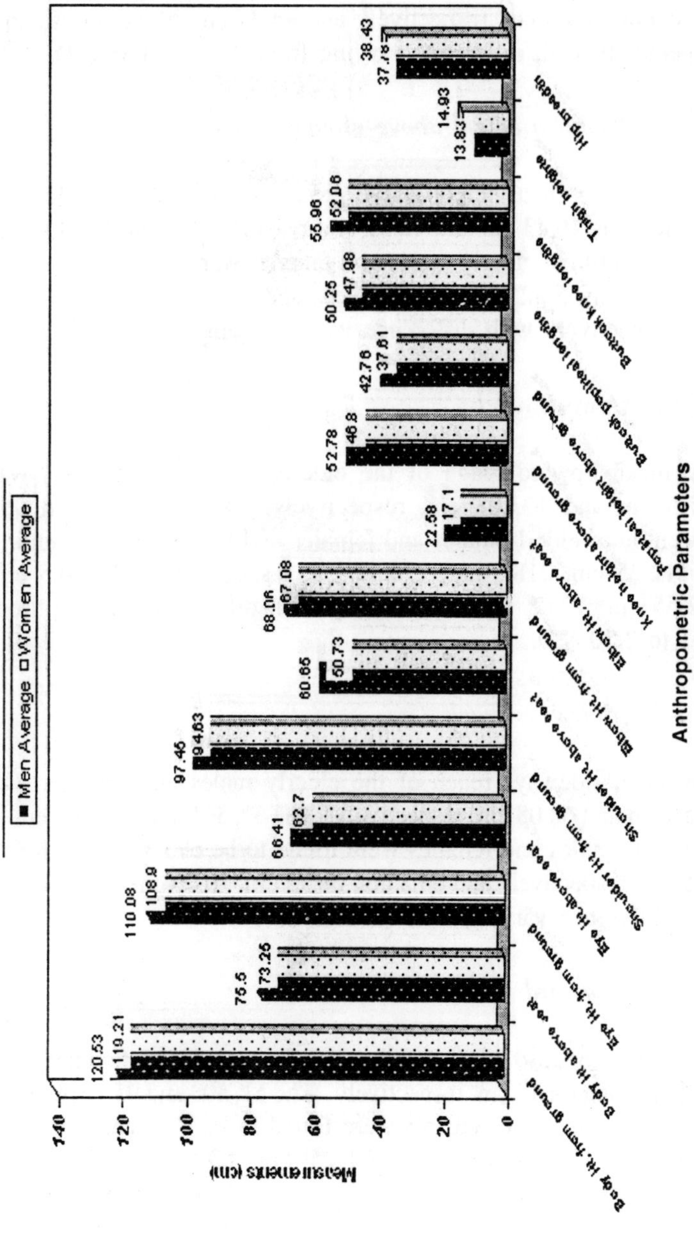

Fig. 4.4(b) : Sex-wise anthropometric measurements (sitting position)

percentile of elderly males and females were found to be 52.45 cm, 57.65 cm, 58.85 cm, respectively and 44.37 cm, 50 cm, 52.97 cm, respectively with difference varying from 2.93 to 3.58 S.D.

4.7.1.8 Popliteal height above ground

The mean popliteal height above ground of the elderly males and females was 46.43 cm and 42.31 cm, respectively. The 5th, 50th and 95th percentile of elderly males and females were found to be 42.85 cm, 46.4 cm, 50.25 cm, respectively and 40.7 cm, 42.4cm, 44.12 cm, respectively with difference varying from 3.01 to 1.41 S.D.

4.7.1.9 Upward reach

The mean upward reach of the elderly males and females was 217.5 cm and 173.08 cm, respectively. The 5th, 50th and 95th percentile of elderly males and females and females were found to be 212.25 cm, 218.5 cm, 220.75 cm, respectively and 170.5 cm, 172.65 cm, 175.8 cm, respectively with difference varying from 3.61 to 2.20 S.D.

4.7.1.10 Sideways reach

The mean sideways reach of the elderly males and females was 71.58 cm and 68.08 cm, respectively. The 5th, 50th and 95th percentile of elderly males and females were found to be 69.07 cm, 71.8 cm, 73.5 cm, respectively and 64.5 cm, 68.75 cm, 70.75 cm, respectively with difference varying from 1.88 to 2.61 S.D.

4.7.1.11 Forward reach

The mean forward reach of the elderly males and females was 74.73 cm and 64.45 cm, respectively. The 5th, 50th and 95th percentile of elderly males and females were found to be 72.8 cm, 74.7 cm, 76.35 cm, respectively and 61.25 cm, 63.25 cm, 70.4 cm, respectively with difference varying from 1.49 to 4.03 S.D. *Stoudt (1981)* noted that functional reach is diminished, at times severely.

4.7.1.12 Shoulder-elbow height

The mean shoulder-elbow height of the elderly males and females was 33.9 cm and 30.38 cm, respectively. The 5th, 50th and 95th percentile of elderly males and females were found to be 32.2 cm, 33.55 cm, 35.87 cm, respectively and 29.3 cm, 30.6 cm, 31.37 cm, respectively with difference varying from 1.58 to 0.90 S.D.

4.7.1.13 Forearm-elbow height

The mean forearm-elbow height of the elderly males and females was 42.85 cm and 38.76 cm, respectively. The 5th, 50th and 95th percentile of elderly males and females were found to be 42.12 cm, 42.95 cm, 43.42 cm, respectively and 38.02 cm, 38.85 cm, 39.45 cm, respectively with difference varying from 0.55 to 0.61 S.D.

4.7.1.14 Hand length

The mean hand length of the elderly males and females was 20.55 cm and 17.35 cm respectively. The 5th, 50th and 95th percentile of elderly males and females were found to be 19.12 cm, 20.25 cm, 22.47 cm, respectively and 15.65 cm, 17.45 cm, 18.75 cm, respectively with difference varying from 1.39 to 1.26 S.D.

4.7.1.15 Elbow to elbow breadth

The mean elbow to elbow breadth of the elderly males and females was 49.41 cm and 43.86 cm, respectively. The 5th, 50th and 95th percentile of elderly males and females were found to be 47.47 cm, 49.75 cm, 50.87 cm, respectively and 41 cm, 43.5 cm, 47.65 cm, respectively with difference varying from 1.42 to 2.98 S.D.

4.7.1.16 Total arm length

The mean total arm length of the elderly males and females was 71.78 cm and 62 cm, respectively. The 5th, 50th and 95th percentile of elderly males and females were found to be 69 cm, 72.2 cm, 73.67 cm, respectively and 56.5 cm, 63.5 cm, 65.75 cm, respectively with difference varying from 2.02 to 4.04 S.D. The findings of correlation co-efficient (*r* value) showed the highly significant relationship between anthropometric measurement (standing posture) of elderly male and female at 1 per cent level of significance.

4.7.2 Anthropometric measurements of elderly people (sitting position)

4.7.2.1 Body height from ground

The mean body height from ground of the elderly males and females was 120.53 cm and 119.21 cm, respectively. The 5th, 50th and 95th percentile of elderly males and females were found to be 118.52 cm, 120.75 cm, 122 cm and 114.5 cm, 120.75 cm, 121.95 cm, respectively with difference varying from 1.90 to 3.37 S.D. Sitting height also declines with age (*Borkan et al. 1983*) in parallel to the decline in stature (*Stoudt, 1981*).

4.7.2.2 Body height above seat

The mean body height above seat of the elderly males and females was 75.5 cm and 73.25 cm respectively. The 5th, 50th and 95th percentile of elderly males and females were found to be 73.82 cm, 75.65 cm, 76.9 cm, respectively and 72.52 cm, 73.35 cm, 73.87 cm, respectively with difference varying from 5.91 to 0.55 S.D.

4.7.2.3 Eye height from ground

The mean eye height from ground of the elderly males and females was 110.08 cm and 108.9 cm respectively. The 5th, 50th and 95th percentile of elderly males and females were found to be 108.3, 110.87 cm, 110.97 cm, respectively and 106.7 cm, 109.75 cm, 110.37 cm, respectively with difference varying from 1.17 to 1.90 S.D.

4.7.2.4 Eye height above seat

The mean eye height above seat of the elderly males and females was 66.41 cm and 62.7 cm, respectively. The 5th, 50th and 95th percentile of elderly males and females were found to be 64.62 cm, 66.45 cm, 68.12 cm, respectively and 61.17 cm, 62.5 cm, 64.45 cm, respectively with difference varying from 1.43 to 1.36 S.D.

4.7.2.5 Shoulder height from ground

The mean shoulder height from ground of the elderly males and females was 97.45 cm and 94.63 cm, respectively. The 5th, 50th and 95th percentile of elderly males and females were found to be 90.5 cm, 99.35 cm, 101.25 cm, respectively and 89.85 cm, 96.2 cm, 98 cm, respectively with difference varying from 4.70 to 3.57 S.D.

4.7.2.6 Shoulder height above seat

The mean shoulder height above seat of the elderly males and females was 60.65 cm and 50.73 cm, respectively. The 5th, 50th and 95th percentile of elderly males and females were found to be 54.12 cm, 61.7 cm, 65.75 cm and 49.42 cm, 50.8 cm, 51.92 cm, respectively with difference varying from 5.14 to 1.05 S.D.

4.7.2.7 Elbow height from ground

The mean elbow height from ground of the elderly males and females was and 67.08 cm, respectively. The 5th, 50th and 95th percentile of elderly males and females were found to be 62.87 cm, 68.95 cm, 70.75 cm, respectively and 65.17 cm, 66.95 cm, 69.32 cm, respectively with difference varying from 3.58 to 1.69 S.D.

4.7.2.8 Elbow height above seat

Elbow rest height is used to determine the height of the arm rests. The mean elbow height above seat of the elderly males and females was 22.58 cm and 18.08 cm, respectively. The 5th, 50th and 95th

percentile of elderly males and females were found to be 21.25 cm, 22.75 cm, 23.75 cm, respectively and 17.02 cm, 18.05 cm, 19.17 cm, respectively with difference varying from 1.02 to 1.08 S.D.

4.7.2.9 Knee height above ground

The mean knee height above ground of the elderly males and females was 52.78 cm and 46.8 cm, respectively. The 5th, 50th and 95th percentile of elderly males and females were found to be 51.25 cm, 53.1 cm, 53.87 cm and 44.12 cm, 46.9 cm, 48.72 cm, respectively with difference varying from 1.11 to 1.93 S.D.

4.7.2.10 Popliteal height above ground

The popliteal height, after adjusting for heels, clothing and other issues is used to determine the height of the chair seat. The mean popliteal height above ground of the elderly males and females was 42.76 cm and 37.61 cm, respectively. The 5th, 50th and 95th percentile of elderly males and females were found to be 40.35 cm, 43.2 cm, 43.95 cm, respectively and 35.75 cm, 37.75 cm, 39.25 cm, respectively with difference varying from 1.65 to 1.44 S.D.

4.7.2.11 Buttock popliteal length

Buttock popliteal length is used to determine the seat depth. The mean buttock popliteal length of the elderly males and females was 50.25 cm and 47.98 cm, respectively. The 5th, 50th and 95th percentile of elderly males and females were found to be 48.52 cm, 50.75 cm, 51.37 cm, respectively and 46.07 cm, 48 cm, 49.65 cm, respectively with difference varying from 1.23 to 1.50 S.D.

4.7.2.12 Buttock knee length

Buttock-knee length is used to determine "leg room" between rows of chairs. The mean buttock knee length of the elderly males and females was 55.96 cm and 52.06 cm, respectively. The 5th, 50th and 95th percentile of elderly males and females were found to be 55.52 cm,

56 cm, 56.37 cm, respectively and 49.22 cm, 52.5 cm, 53.62 cm, respectively with difference varying from 0.35 to 1.95 S.D.

4.7.2.13 Thigh height

The mean thigh height of the elderly males and females was 13.83 cm and 14.93 cm, respectively. The 5th, 50th and 95th percentile of elderly males and females were found to be 11.8 cm, 14.3 cm, 15.15 cm, respectively and 14.12 cm, 15.1 cm, 15.52 cm, respectively with difference varying from 1.46 to 0.58 S.D.

4.7.2.14 Hip breadth

Hip breadth is used for chair width and arm rest width. The mean hip breadth of the elderly males and females was 37.78 cm and 38.43 cm, respectively. The 5th, 50th and 95th percentile of elderly males and females were found to be 33.75 cm, 38.65 cm, 40.4 cm, respectively and 34.27 cm, 39.2 cm, 40.57 cm, respectively; with difference varying from 2.83 to 2.83 S.D. The findings of correlation co-efficient showed the highly significant relationship from anthropometric measurement (sitting posture) of elderly male and female at 1 per cent level of significance.

The most critical dimensions for furniture design are seat height and table height and according to the *Evans et al. (1988)* seat height should correspond to the 5th percentile popliteal height of target population with an adjustment for shoes. Data collected in present study showed that the 5th percentile value of popliteal height was 40.35 with 2.5 cm of allowance, *i.e.* 42.85 cm but seat height of existing chair was 45.04 cm. It showed that furniture is not ergonomically sound and did not match with the measurement of elderly people.

4.8 Environmental parameters in old age homes

This part comprise of findings pertaining to the measurement of four micro-environment which refers to our immediate surrounds with which we closely interact, *i.e.* light, sound, humidity and temperature in the room. Ideally, the environment provides more than shelter. It should promote continued development, stimulation and satisfaction to enhance our psychological well being. This is particularly important for the elderly people, many of whom spend considerable time in their rooms of old age homes.

Table 4.8 : Environmental parameters in old age homes

Environmental parameters	*Old age home*					
	I	*II*	*III*	*IV*	*Mean*	*S.D.*
1. Relative humidity (%)	68	69	70	64	67.75	2.62
2. Ambient temperature (°C)	25.7	30.4	26.5	27.8	27.6	2.05
3. Noise level (dB)	39.1	46.3	40.1	42.3	41.95	3.19
4. Light level (Lux)						
(a) Habitat room	45	31.8	32.9	34	35.92	6.11
(b) Bathroom	25	17	20.5	23	21.37	3.44

4.8.1 Relative humidity

Relative humidity is usually measured as a percentage of moisture in air in relation to the moisture that would completely saturate the air at the existing temperature. The relative humidity should not normally exceed 70 per cent *(Galar, 1995).*

The average humidity level of four old age homes was 67.75 per cent; with 2.62 S.D. falls within the recommended value. A comparative study of relative humidity can be made under the following categories:

0-25 %	Very dry
25-50 %	Dry
50-75 %	Humid
75-100 %	Very humid

Accordingly the humidity level in all the old age homes falls in humid range category. *Harold (2000)* mentioned an ideal indoor relative humidity from 30 per cent and 60 per cent to minimize growth of allergenic or pathogenic organisms.

4.8.2 Ambient temperature

Housing must provide protection from extremes of temperature as well as other environmental hazards. Elderly people are sensitive to lower environmental temperatures because of their lower body temperature and decreased amount of natural insulation make the elderly people especially sensitive to lower temperature. The average temperature in all the four old age homes was 27.6ºC in the month of August 2006. The recommended room temperature for an older person should not be lower than 75ºF (24ºC). Room temperature less than 70ºF (21ºC) can lead to hypothermia in the elderly people. Brain damage can result from temperature exceeding 106ºF (41ºC). The temperatures categories are as under:

Below 15°C	Very cold
15°C to 20°C	Cool
20°C to 30°C	Optimum or comfortable
30°C to 35°C	Hot
Above 35°C	Very hot

Comfortable body temperature is 20˚C and 23˚C; it showed that the temperatures in the room of all the old age homes were not comfortable.

4.8.3 Sound level

Sound produces a variety of physiologic and emotional effects. Many of the sounds we take for granted—television, traffic noise, conversation from an adjoining room, appliance motors, leaking faucets can create difficulties for the elderly people. Many elderly already experience some hearing limitation as a result of prebycusis and need to be especially attentive to compensate for this defect. Environments noise can have a profound effect on quality of life and some circumstances can lead to hypertension. The average noise level in all the old age homes was 41.95 dB with 3.19 S.D. The recommended value for room is 30 to 45 dB *(WHO)*. In present study all the rooms of old age homes were found within the recommended values of noise level.

4.8.4 Light level

As the visual system ages, a number of changes in its structure and capabilities occur. The consequences of these changes with age are reduced visual acuity, reduced contrast sensitivity, reduced colour discrimination, increased time taken to adapt to large and sudden changes in adaptation luminescence and increased sensitivity to glare (*Boyce, 2003*). The average illumination levels of four old age homes in the habitat room and bathroom were 35.92 Lux; with 6.11 S.D. and 21.37 Lux; with 3.44 S.D. was found to be far less than the prescribed Indian electrical standard, *i.e.* 50 Lux for habitat room and 100 Lux for bathroom. This showed that the old age homes provided insufficient illumination level to the elderly people may make elderly people prone to accidents in area like bathroom etc.

4.9 Postural stress

Haslegrave (1994) reported that posture is as important for the performance of tasks as it is for promoting health and minimizing stress and discomfort during work. Posture arises from the functional demands of vision, reach, manipulation, strength and endurance and is constrained by the geometric relationship between the person's own anthropometric and the layout of workplace.

4.9.1 Upper extremities shoulder external rotation/flexion

Hand and arm posture while using upper shelves of the storage spaces is frequently mentioned as a risk factor for upper extremity musculoskeletal disorders. From Table 4.9.1 it was found that in shoulder external rotation, in using electric switch board required highest average total range of motion value of 62° followed by opening/closing door latches (43.6°), using platform for keeping articles (39.3°), and minimum for opening/closing window (25.6°). In using electric switch board was most stressful one compare to the other activities. In shoulder flexion position, picking articles from floor required highest average total range of motion value of

Table 4.9.1 : Upper extremities shoulder external rotation/flexion

Type of activity	*Primary Sensor (°) Mean Repetitions*			*Total Range of Motion Mean (°)*	*Deviation (360°-ROM Value)*
	Ist	*IInd*	*IIIrd*		
External Rotation					
1. Opening/closing door latches	45°	44°	42°	43.6°	316.4°
2. Opening/closing window	23°	29°	25°	25.6°	334.4°
3. Using electric switch board	62°	64°	60°	62°	298°
4. Using platform for keeping articles	43°	38°	37°	39.3°	320.7°
5. Using T.V./music system	36°	39°	41°	38.6°	321.4°
Shoulder Flexion					
6. Using platform for keeping articles	58°	47°	55°	53.3°	306.7°
7. Using lower shelve of cupboard	65°	75°	72°	70.6°	289.4°
8. Picking articles from floor	105°	105°	109°	106.3°	253.7°

106.3° followed by using lower shelve of cupboard (70.6°) and minimum for using platform for keeping articles (53.3°). It was observed that the picking articles from floor were most stressful one compared to the other activities. Hence, the articles should be kept within comfortable reach. The normal range of shoulder external rotation = zero to 90 degree and shoulder forward flexion = zero to 180 degree and from the findings it was found in normal ranges. *Grozdanovic (2002)* noted that "muscular loading during upper limb intensive work has been linked to the development of chronic muscle problems in the shoulder..." Two common stresses to shoulders are extreme shoulder flexion that occurs in long reaches, and static shoulder loading that occurs when people have to work with their arms elevated.

4.9.2 Upper extremities cervical flexion/extension

Of the various anatomical structures, the cervical muscles and filamentous structures are very susceptible to injury. A forward head position and rounded shoulders have been implicated in the

Table 4.9.2 : Upper extremities cervical flexion/extension

Type of activity	*Primary Sensor (°)*		*Secondary Sensor (°)*		*The difference in angle (°) Total ROM*		*Deviation (180°- The difference in angle ROM Value)*
	Repetitions	*ROM Value*	*Repetitions*	*ROM Value*	*Repetitions*	*ROM Value*	
Cervical Flexion Picking articles from cupboard (lower shelve)	Ist	38°	Ist	28°	Ist	10°	170°
	IInd	36°	IInd	27°	IInd	9°	171°
	IIIrd	44°	IIIrd	39°	IIIrd	5°	175°
Mean		**39.3°**		**31.3°**		**8°**	**172°**
Cervical Extension Picking articles from cupboard (upper shelve)	Ist	30°	Ist	1°	Ist	29°	151°
	IInd	25°	IInd	1°	IInd	24°	156°
	IIIrd	25°	IIIrd	3°	IIIrd	22°°	158°
Mean		**26.6°**		**1.6°**		**25°**	**155°**

development or perpetuation of craniomandibular disorders. The findings summarized in Table 4.9.2 exhibited that in cervical flexion and cervical extension, the average Primary Sensor (Rom Value) was 39.3° and 26.6°. However, in cervical flexion movement the average of secondary sensor ROM value of 31.3° was greater than in cervical extension 1.6°, while the difference between (Primary sensor value – Secondary sensor value), *i.e.* total range of motion value (25°) was greater in cervical extension than cervical flexion (8°). The findings depicted that in cervical extension movement (picking article from cupboard upper shelve) was more stressful than in cervical flexion movement (picking article from cupboard lower shelve). The normal range of cervical flexion/extension = zero to 45 degree and from the findings it was found in normal ranges.

4.9.3 Gross spinal flexion range of motion

Spinal flexion was measured for using storage spaces of the lower and the middle shelves in the room. According to *Barton et al. (1992)*, by the age of 65, 90 per cent of the population has radiological evidence of cervical spondolysis due to poor posture at work. Cervical spine degradation is a potential cause of neck pain because of the mechanical changes, which occur as a result of age related degenerative processes.

It was found that the average of Primary Sensor and Secondary Sensor in using lower shelves and middle shelves was 136.96°, 132°, and 112.3°, 110.43°, respectively. The average of the difference in angle, *i.e.* Total Range of Motion 5.05° was greater in using storage spaces—lower shelves while using storage spaces—middle shelves (1.87°). It revealed that the using lower shelves were much more stressful than that of using storage spaces—middle shelves. Neglect of ergonomic principles brings inefficiency and pain to the elderly people in workplace.

4.10 Assessment of pain perceived by the elderly people

Improvements in workplace, working posture, and discomfort need to be justified in terms of improvements in performance. Corelett

Table 4.9.3 : Gross spinal flexion range of motion while using storage spaces

Type of activity	*Repetitions*	*Primary Sensor (°)*	*Repetitions*	*Secondary Sensor (°)*	*Repetitions*	*The difference in angle Total ROM Value (°)*	*Deviation (180°— The difference in angle ROM Value)*
Using storage spaces (lower shelve)	1	133.3°	1	129.0°	1	4.3°	175.7°
	2	137.3°	2	132.3°	2	5.8°	174.2°
	3	135.8°	3	130.1°	3	5.6°	174.4°
	4	137.1°	4	133.1°	4	4.0°	174°
	5	137.8°	5	133.5°	5	4.5°	175.5°
	6	140.5°	6	134°	6	6.1°	173.9°
Mean		**136. 96°**		**132°**		**5.05°**	**174.95°**
Using storage spaces (Middle shelve)	1	121.1°	1	119.3°	1	1.8°	178.2
	2	121°	2	119.5°	2	1.5	178.5
	3	111.1°	3	109.1°	3	2°	178°
	4	111.3°	4	109.6°	4	1. 7°	178.3°
	5	102°	5	99.8°	5	2.2°	177.8°
	6	107.3°	6	105.3	6	2°	178°
Mean		**112. 3°**		**110. 43°**		**1. 87°**	**178.13°**

and Bishop CRS ten points rating technique *(Noble et al. 1983)* was used for pain perceived by elderly people related to different body parts like neck, shoulder, arms, hands, knees, legs, buttock and back.

Genaidy and Karwowski (1993) investigated the discomfort associated with joint postures, which deviated from the natural position. It was revealed that in standing, lumber extension was perceived to be more uncomfortable than lateral lumber bending or rotation. In both standing and sitting, flexion of the back was the least uncomfortable of the extreme postures.

4.10.1 Rating of intensity of pain

The selected elderly people were asked to answer a pain questionnaire and to stand by a plumb line for pain perceived.

4.10.1.1 While using chair

The findings summarized in t-table 4.10.1 showed that while using furniture, *i.e.* chair and cot, window, ventilator and storage spaces (upper and lower shelves), elderly people responded up to five points in rating scale of intensity of pain. In old age home Ist, mid back pain was found to be a major pain area while using chair with mean score of 54, whereas legs, buttocks were also rated as major discomfort and pain areas with the total score of 52, 46 respectively. The reason behind this was absence of foot rest and cushions in seat and back of the chair. In old age home IInd, buttock pain was found to be most prevalent and somewhat strong intensity of pain as majority of elderly people rated up to four points with total score 56 and for legs and back 55, 50 respectively. In old age home IIIrd, legs pain was found to be more severe in comparison to other body parts with mean score of 3.2, followed by buttock 3.1 and mid back 2.9, while lowest for shoulder 2.4. In old age home IVth back pain was found to be more severe with highest 61 score followed 55 in thigh, and lowest 30 in shoulder. Total score was 335 with mean score of 22.3. The grand total score obtained was 1291 with mean score of 21.4 from all the old age homes (Fig.4.5a).

Table 4.10.1 : Distribution of elderly people by their intensity of pain while using chair

Old age home	Body parts	Rating of Intensity of Pain											Total score	Mean score
		0	0.5	1	2	3	4	5	6	7	8 & 9	10		
		Nothing at all	*Extremely weak*	*Very weak*	*Weak*	*Moderate*	*Some-what strong*	*Strong*	*In between strong and v. strong*	*Very strong*	*In between strong & extremely strong*	*Extremely strong*		
I.	Mid back	—	—	—	—	6 (40)	9 (60)	—	—	—	—	—	54	3.6
	Buttock	—	—	—	—	2 (13)	10 (66.6)	3 (20)	—	—	—	—	46	3.0
	Legs	—	—	—	—	8 (53)	7 (46)	—	—	—	—	—	52	3.4
	Thighs	—	—	—	—	12 (80)	3 (20)	—	—	—	—	—	48	3.2
	Arms	—	—	—	10 (66)	5 (33.3)	—	—	—	—	—	—	35	2.3
	Neck	—	—	—	—	9 (60)	6 (40)	—	—	—	—	—	51	3.4
	Shoulder	—	—	—	15 (100)	—	—	—	—	—	—	—	30	2.0
	Total												**316**	21.0
II.	Mid back	—	—	—	—	10 (66)	5 (33)	—	—	—	—	—	50	3.3
	Buttock	—	—	—	—	4 (26)	11 (73)	—	—	—	—	--	56	3.7
	Legs	—	—	—	—	5 (33)	10 (66.6)	—	—	—	—	—	55	3.6
	Thighs	—	—	—	—	12 (80)	3 (20)	—	—	—	—	—	48	3.2
	Arms	—	—	—	2(13.3)	13 (86)	—	—	—	—	—	—	43	2.8
	Neck	—	—	—	—	8 (53)	7 (46)	—	—	—	—	—	52	3.4
	Shoulder	—	—	—	7 (46)	8 (53)	—	—	—	—	—	—	38	2.5
	Total												**342**	**22.8**

(Contd.)

Table 4.10.1. : (Contd.)

III.	Mid back	—	—	—	4 (26.6)	8 (53)	3 (20)	—	—	—	—	—	44	2.9
	Buttock	—	—	—	3 (20)	7 (46)	5 (33)	—	—	—	—	—	47	3.1
	Legs	—	—	—	1 (6.6)	10 (66)	4 (26)	—	—	—	—	—	48	3.2
	Thighs	—	—	—	7 (46)	6 (40)	2 (13.3)	—	—	—	—	—	40	2.6
	Arms	—	—	—	4 (26)	10 (66)	1 (6.6)	—	—	—	—	—	42	2.8
	Neck	—	—	—	7 (46)	6 (40)	2 (13.3)	—	—	—	—	—	40	2.6
	Shoulder	—	—	—	8 (53)	7 (46)	—	—	—	—	—	—	37	2.4
	Total												**298**	**19.8**
IV.	Mid back	—	—	—	—	2 (13)	10 (66.6)	3 (20)	—	—	—	—	61	4.0
	Buttock	—	—	—	—	8 (53)	7 (46)	—	—	—	—	—	52	3.4
	Legs	—	—	—	—	12 (80)	3 (20)	—	—	—	—	—	48	3.2
	Thighs	—	—	—	—	5 (33)	10 (66.6)	—	—	—	—	—	55	3.6
	Arms	—	—	—	10 (66.6)	5 (33)	—	—	—	—	—	—	35	2.3
	Neck	—	—	—	—	6 (40)	9 (60)	—	—	—	—	—	54	3.6
	Shoulder	—	—	—	15 (100)	—	—	—	—	—	—	—	30	20
	Total												**335**	**22.3**
	Grand Total												**1291**	**21.4**

Note : Figures in parentheses represent percentage.

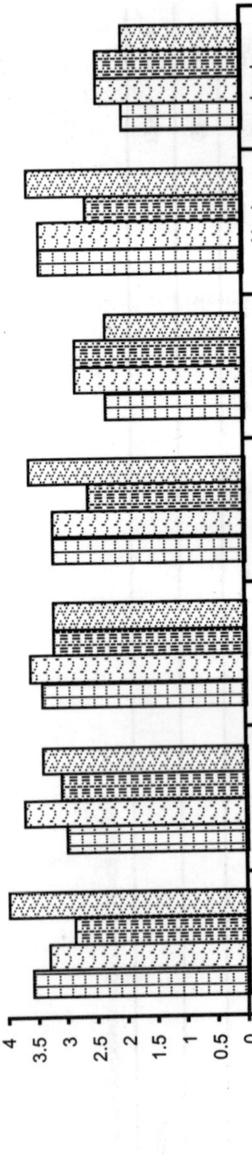

	Back	Buttock	Legs	Thighs	Arms	Neck	shouler
Old age home I	3.6	3	3.4	3.2	2.3	3.4	2
Old age home II	3.3	3.7	3.6	3.2	2.8	3.4	2.4
Old age home III	2.9	3.1	3.2	2.6	2.8	2.6	2.4
Old age home IV	4	3.4	3.2	3.6	2.3	3.6	2

Fig. 4.5 (a): Distribution of elderly people by their scores on perceived pain while using chair

4.10.1.2 While using cot

The findings showed that in all the four old age homes, the elderly people rated up to five points in pain intensity scale. This was one point higher than chair users, because the cots used for sleeping as well as sitting purpose by the elderly people in most of the times. In old age home Ist, while sitting on cot, almost all the elderly people felt severe discomfort and somewhat strong pain in legs and thighs with mean score of 4.0, whereas lowest 1.3 for buttock pain. The reason behind this pain was the height of cot from floor. Total score obtained was 160 lowest compared to other old age homes. In old age home IInd, the elderly people felt discomfort and pain in legs and thighs while sitting and sleeping on cot, with mean score of 4.0 and 3.6, respectively and lowest for back. In old age home IIIrd, 26 per cent, 53 per cent and 20 per cent of elderly people felt moderate, somewhat strong and strong intensity of pain in back, respectively, while sleeping on cot. The reason behind this was that the cots were made up of metal and the mattresses used were thin. The highest mean score of 4.6 was recorded for legs and lowest 3.6 for thighs. Total score obtained was highest, *i.e.* 242 in comparison to other old age homes. In old age home IVth the highest score 4.0 was observed for back pain and lowest 3.0 in thighs while using cots for sitting and sleeping purposes. The grand total score obtained from all the old age homes was 746 with mean score of 12.4 (Fig. 4.5b).

4.10.1.3 While using window

The result showed that in old age home Ist while using window, 33, 80, and 13 per cent of elderly people felt moderate intensity of pain, while 66, 20, and 86 per cent somewhat strong intensity of pain in their arms, neck and shoulders, respectively. Total score obtained was 161 with highest mean score of 3.8 for shoulder pain. In old age homes IInd and IIIrd, 46, 86 and 40 per cent and 66, 86 and 33 per cent of the elderly people felt moderate intensity of pain in arms, neck and shoulder, respectively, while 53, 13, 60 per cent and 33, 13 and 40 per cent felt somewhat strong intensity of pain in arms, neck and shoulder, respectively. In old age home IVth 46,

Table 4.10.2 : Distribution of elderly people by their intensity of pain while using cot

Old age home	Body parts	Body parts Rating of Intensity of Pain											Total score	Mean score
		0	0.5	1	2	3	4	5	6	7	8 & 9	10		
		Nothing at all	Extremely weak	Very weak	Weak	Moderate	Some-what strong	Strong	In between strong and v. strong	Very strong	In between strong & extremely strong	Extremely strong		
I.	Buttock	—	—	—	10(66)	—	5(33)	—	—	—	—	—	40	2.6
	Legs	—	—	—	—	—	15(100)	—	—	—	—	—	60	4.0
	Thigh	—	—	—	—	—	12(80)	3(20)	—	—	—	—	60	4.0
	Total												**160**	**10.6**
II.	Back	—	—	—	12(80)	3(20)	—	—	—	—	—	—	33	2.2
	Buttock	—	—	—	9 (60)	6 (40)	—	—	—	—	—	—	36	2.4
	Legs	—	—	—	—	—	15 (100)	—	—	—	—	—	60	4.0
	Thigh	—	—	—	—	5(33)	10(66)	—	—	—	—	—	55	3.6
	Total												**184**	**12.2**
III.	Back	—	—	—	—	4 (26)	8 (53)	3 (20)	—	—	—	—	59	3.9
	Buttock	—	—	—	—	3 (20)	12 (80)	—	—	—	—	—	54	3.6
	Legs	—	—	—	—	6 (40)	9 (60)	—	—	—	—	—	69	4.6
	Thigh	—	—	—	—	—	15 (100)	—	—	—	—	—	60	4.0
	Total												**242**	**16.1**

(Contd.)

Table 4.10.2 : (Contd.)

IV	Back	—	—	—	—	3 (20)	9 (60)	3 (20)	—	—	—	—	60	4.0
	Buttock	—	—	—	—	5 (33)	10 (66)		—	—	—	—	55	3.6
	Legs	—	—	—	—	2 (13)	13 (86)		—	—	—	—	45	3.0
	Total												**160**	**10.6**
	Grand Total												**746**	**12.4**

Note : Figures in parentheses represent percentage.

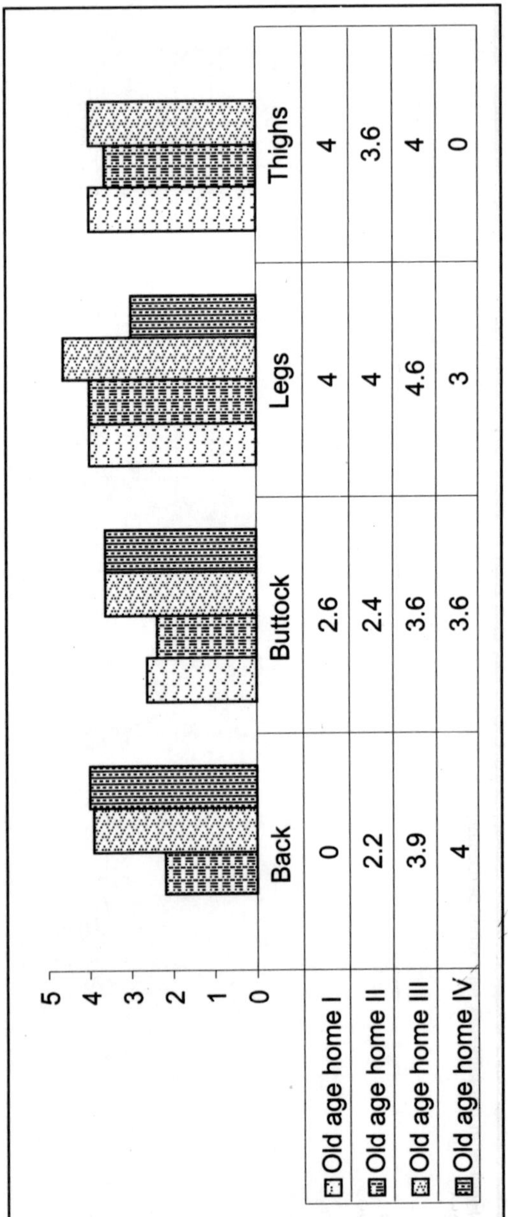

Fig. 4.5(b): Distribution of elderly people by their scores on perceived pain while using cot

Table 4.10.3 : Distribution of elderly people by their intensity of pain while using window

Old age home	Body parts	Rating of Intensity of Pain											Total score	Mean score
		0 Nothing at all	*0.5 Extremely weak*	*1 Very weak*	*2 Weak*	*3 Moderate*	*4 Some what strong*	*5 Strong*	*6 In between strong and v. strong*	*7 Very strong*	*8 & 9 In between strong & extremely strong*	*10 Extremely strong*		
I.	Arms	—	—	—	—	5 (33)	10 (66)	—	—	—	—	—	55	3.6
	Neck	—	—	—	—	12 (80)	3 (20)	—	—	—	—	—	48	3.2
	Shoulder	—	—	—	—	2 (13)	13 (86)	—	—	—	—	—	58	3.8
	Total												**161**	**10.7**
II.	Arms	—	—	—	—	7 (46)	8 (53)	—	—	—	—	—	53	3.5
	Neck	—	—	—	—	13 (86)	2 (13)	—	—	—	—	—	47	3.1
	Shoulder	—	—	—	—	6 (40)	9 (60)	—	—	—	—	—	54	3.6
	Total												**154**	**10**
III.	Arms	—	—	—	—	10 (66)	5 (33)	—	—	—	—	—	50	3.3
	Neck	—	—	—	—	13 (86)	2 (13)	—	—	—	—	—	47	3.1
	Shoulder	—	—	—	—	8 (53)	7 (40)	—	—	—	—	—	52	3.4
	Total												**149**	**9.9**
IV.	Arms	—	—	—	—	7 (46)	8 (53)	—	—	—	—	—	53	3.5
	Neck	—	—	—	—	9 (60)	6 (40)	—	—	—	—	—	59	3.9
	Shoulder	—	—	—	—	2 (13)	13 (86)	—	—	—	—	—	58	3.8
	Total												**170**	**11.3**
	Grand Total												**634**	**10.5**

Note : Figures in parentheses represent percentage.

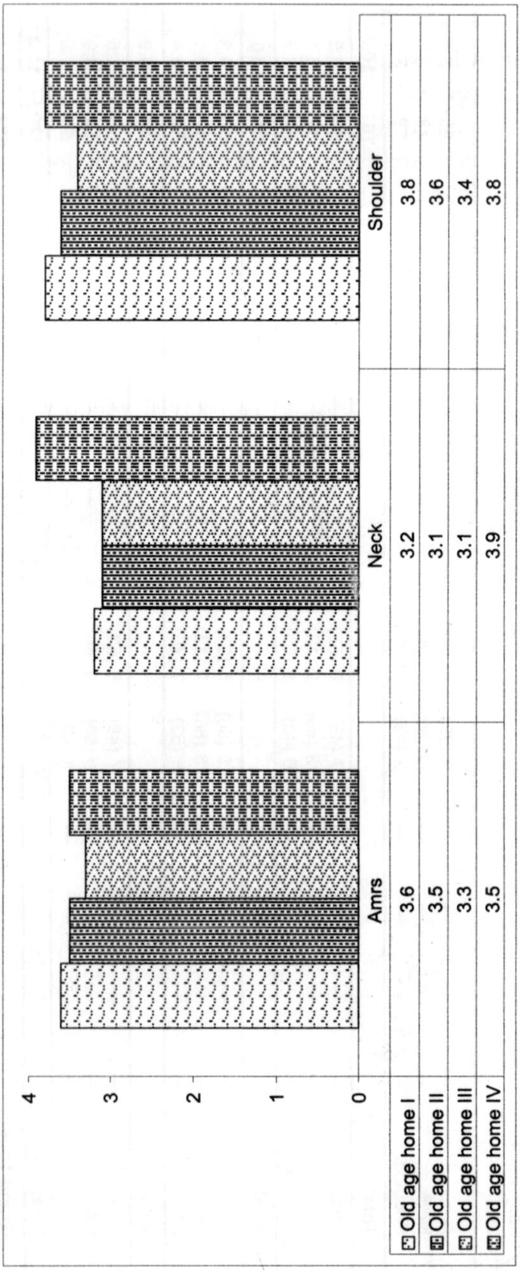

Fig. 4.5(c): Distribution of elderly people by their scores on perceived pain while using window

60, 13 per cent of elderly people felt moderate, while 53, 40, 86 per cent somewhat strong intensity of pain in their arms, neck, and shoulders, respectively. Total score obtained was 170, which was highest as compared to other old age homes. The reason behind this was the increased height of windows, as majority of elderly females were found facing difficulties to open/close it. The grand total score from all the old age home was 634 with mean score of 10.5 (Fig.4.5c).

4.10.1.4 While using ventilator

Elderly people responded up to five points, which was one point higher than while using windows. In old age home I[st], 86 and 13 per cent of elderly people felt moderate and somewhat strong intensity of pain in legs, while 66 and 33 per cent in arms, respectively. Sixty-six, 33 per cent felt somewhat strong and strong intensity of pain in shoulder. The total score obtained was 207 with highest mean score of 4.3 in shoulder. Whereas, in old age home IInd, 73, 20, and 66 per cent of elderly people felt moderate, somewhat strong, and strong intensity of pain in arms, respectively, while 33 and 66 per cent felt somewhat strong, and strong intensity of pain in shoulder and almost all felt moderate pain in neck in old age homes I[st] , IInd, IIIrd, and IVth. The total score obtained was 165 with mean score of 4.6 highest for shoulder. In old age home IIIrd, 73, 20 and 66 per cent of elderly people felt moderate, somewhat strong and strong pain in their legs, respectively, while 26, 60, 13 per cent felt moderate, somewhat strong, and strong intensity of pain in their arms, respectively, while 20, 60, 20 per cent felt moderate, somewhat strong, and strong intensity of pain in shoulder, respectively. In old age home IVth, 73 and 26 per cent of elderly people felt moderate and somewhat strong intensity of pain in legs, respectively, while 13, 60 and 26 per cent felt moderate, some what strong and strong intensity of pain in their arm. 55 and 33 per cent felt some whatstrong, strong intensity of pain in shoulder. The total score was highest (219) in old age home IVth, compared to other old age homes, as majority of them were unable to open/ close the ventilators themselves and needed someone's help because

Table 4.10.4 : Distribution of elderly people by their intensity of pain while using ventilator

Old age home / Body parts	Rating of Intensity of Pain											Total score	Mean score
	0 Nothing at all	*0.5 Extremely weak*	*1 Very weak*	*2 Weak*	*3 Moderate*	*4 Some-what strong*	*5 Strong*	*6 In between strong and v. strong*	*7 Very strong*	*8 & 9 In between strong & extremely strong*	*10 Extremely strong*		
I. Legs	—	—	—	—	13 (86)	2 (13)	—	—	—	—	—	47	3.1
Arm	—	—	—	—	10 (66)	5 (33)	—	—	—	—	—	50	3.3
Neck	—	—	—	—	15 (100)	—	—	—	—	—	—	45	3.0
Shoulder	—	—	—	—	—	10 (66)	5 (33)		—	—	—	65	4.3
Total												**207**	**13.8**
II. Arm	—	—	—	—	11 (73)	3 (20)	1 (6.6)	—	—	—	—	50	3.3
Neck	—	—	—	—	15 (100)	—	—	—	—	—	—	45	3.0
Shoulder	—	—	—	—	—	5 (33)	10 (66)	—	—	—	—	70	4.6
Total												**165**	**11.0**
III. Legs	—	—	—	—	11 (73)	3 (20)	1 (6.6)	—	—	—	—	50	3.3
Arm	—	—	—	—	4 (26)	9 (60)	2 (13)	—	—	—	—	58	3.8
Neck	—	—	—	—	15 (100)	—	—	—	—	—	—	45	3.0
Shoulder	—	—	—	—	3 (20)	9 (60)	3 (20)	—	—	—	—	60	4.0
Total												**213**	**14.2**

(Contd.)

Table 4.10.4 : (Contd.)

IV	Legs	—	—	—	—	11(73)	4 (26)	—	—	—	—	—	49	3.2
	Arm	—	—	—	—	2 (13)	9 (60)	4 (26)	—	—	—	—	62	4.1
	Neck	—	—	—	—	15 (100)	—	—	—	—	—	—	45	3.0
	Shoulder	—	—	—	—	2 (13)	8 (53)	5 (33)	—	—	—	—	63	4.2
	Total												**219**	**14.6**
	Grand total												**804**	**13.4**

Note : Figures in parentheses represent percentage.

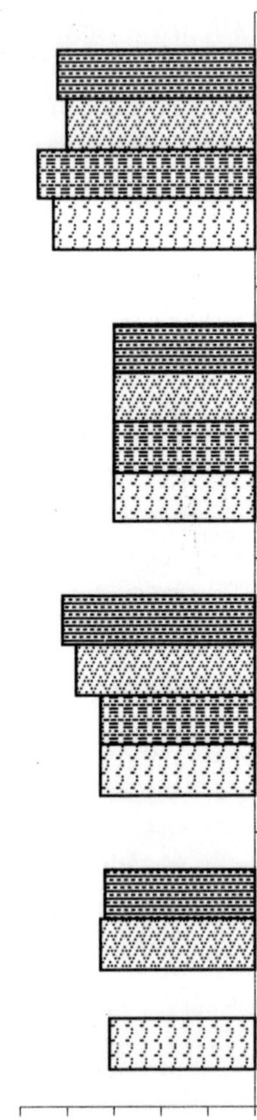

	Leg	Arm	Neck	Shoulder
Old age home I	3.1	3.3	3	4.3
Old age home II	0	3.3	3	4.6
Old age home III	3.3	3.8	3	4
Old age home IV	3.2	4.1	3	4.2

Fig. 4.5(d): Distribution of elderly people by their scores on perceived pain while using ventilator

of the extra increased height of ventilators, which was even out of reach of the elderly females. The grand total score was 804 with mean score of 13.4 from all the old age homes (Fig.4.5d).

4.10.1.5 While using storage space (upper shelves)

The elder people responded up to five points in the pain intensity scale. The findings showed that in old age home I[st], 33 and 66 per cent of elderly people felt somewhat strong and strong intensity of pain in their neck, while 53 and 46 per cent in shoulder, and 60 and 40 per cent in arms, respectively Almost all the elderly people felt moderate pain in hands. The total score was 248; with highest mean score of 4.6 for neck pain, due to the extra height of the upper shelve as they had to stretch their shoulders and arms and needed stool to reach. In old age home IInd, 20 and 80 per cent of elderly people felt somewhat strong and strong pain in their neck, while 33.3 and 53 per cent in shoulder, 60 and 40 per cent in arms, 35 and 66 per cent in hands, and 26 and 73 per cent in legs, respectively. The total score was highest (347) as compared to old age home I[st]. In old age home IIIrd, 66, each 46 per cent, 6.6, 60, 33 per cent, 13.3, 46, 40 per cent and 6.6, 46, 40 per cent of elderly people felt moderate, somewhat strong, and strong intensity of pain in their neck, arms, hands and legs, respectively with the total score of 325. In old age home IVth, 13.3, 20, and 66 per cent of elderly people felt moderate, somewhat strong, and strong intensity of pain in neck, while 13.3 and 86 per cent in shoulders, 66, 13.3 and 80 per cent in arms and 20, 13.3, and 66 per cent in hands, and 6.6, 60, and 33 per cent in legs, respectively. The total score 371 was highest as compared to other old age homes; with highest mean score of 6.4 for neck pain. The grand total score from all the four old age homes was 1291 with mean score of 21.7 (Fig.4.5e).

4.10.1.6 While using storage space (lower shelves)

The elderly people responded up to five points in rating scale in all the old age homes. In old age home I[st], neck pain, while in IInd and

Table 4.10.5 : Distribution of elderly people by their intensity of pain while using cupboard (upper shelves)

Old age home	Body parts	Rating of Intensity of Pain											Total score	Mean score
		0 Nothing at all	0.5 Extremely weak	1 Very weak	2 Weak	3 Moderate	4 Some-what strong	5 Strong	6 In between strong and v. strong	7 Very strong	8 & 9 In between strong & extremely strong	10 Extremely strong		
I.	Neck	—	—	—	—	—	5(33)	10(66)	—	—	—	—	70	4.6
	Shoulder	—	—	—	—	—	8(53)	7(46)	—	—	—	—	67	4.4
	Arms	—	—	—	—	—	9(60)	6(40)	—	—	—	—	66	4.4
	Hand	—	—	—	—	15(100)	—	—	—	—	—	—	45	3.0
	Total												**248**	**16.4**
II.	Neck	—	—	—	—	—	3(20)	12(80)	—	—	—	—	72	4.8
	Shoulder	—	—	—	—	—	7(46)	8(53)	—	—	—	—	68	4.4
	Arms	—	—	—	—	—	9(60)	6(40)	—	—	—	—	66	4.4
	Hand	—	—	—	—	—	5(33)	10(66)	—	—	—	—	70	4.6
	Legs	—	—	—	—	—	4(26)	11(73)	—	—	—	—	71	4.7
	Total												**347**	**23**
III.	Neck	—	—	—	—	1(6.6)	7(46)	7(46)	—	—	—	—	66	4.4
	Shoulder	—	—	—	—	—	10(66)	5(33)	—	—	—	—	65	4.3
	Arms	—	—	—	—	1(6.6)	9(60)	5(33)	—	—	—	—	64	4.2
	Hand	—	—	—	—	2(13.3)	7(46)	6(40)	—	—	—	—	64	4.2
	Legs	—	—	—	—	1(6.6)	7(46)	7(40)	—	—	—	—	66	4.4
	Total												**325**	**21.6**

(Contd.)

Table 4.10.5 : (Contd.)

IV.	Neck	—	—	—	—	2 (13.3)	3 (20)	10 (66)	—	—	—	—	96	6.4
	Shoulder	—	—	—	—	—	2 (13)	13 (86)	—	—	—	—	73	6.8
	Arms	—	—	—	—	1(6.6)	2 (13)	12 (80)	—	—	—	—	71	4.7
	Hand	—	—	—	—	3(20)	2 (13)	10 (66)	—	—	—	—	67	4.4
	Legs	—	—	—	—	1(6.6)	9 (60)	5 (33)	—	—	—	—	64	4.2
	Total												**371**	**24.7**
	Grand Total												**1291**	**21.7**

Note : Figures in parentheses represent percentage.

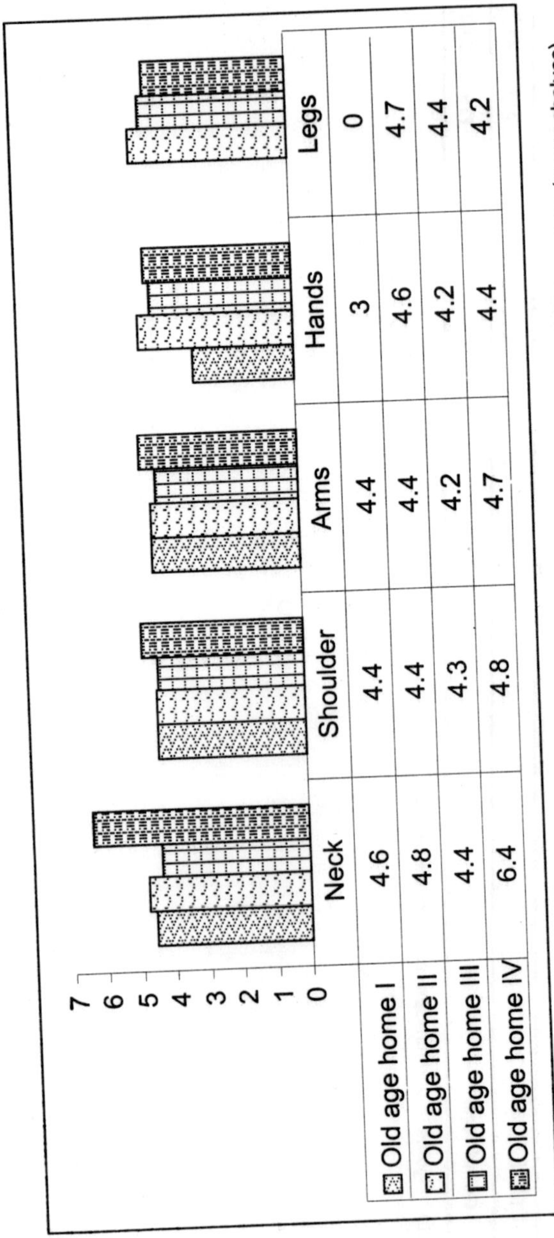

	Neck	Shoulder	Arms	Hands	Legs
Old age home I	4.6	4.4	4.4	3	0
Old age home II	4.8	4.4	4.4	4.6	4.7
Old age home III	4.4	4.3	4.2	4.2	4.4
Old age home IV	6.4	4.8	4.7	4.4	4.2

Fig. 4.5(e) : Distribution of elderly people by their scores on perceived pain while using storage spaces (upper shelves)

Table 4.10.6 : Distribution of elderly people by their intensity of pain while using cupboard (lower shelves)

Old age home	Body parts	0 Nothing at all	0.5 Extremely weak	1 Very weak	2 Weak	3 Moderate	4 Some-what strong	5 Strong	6 In between strong and v. strong	7 Very strong	8 & 9 In between strong & extremely strong	10 Extremely strong	Total score	Mean score
		Rating of Intensity of Pain												
I.	Back	—	—	—	—	4 (36)	9 (60)	2 (33)	—	—	—	—	58	3.8
	Neck	—	—	—	—	2 (13)	5 (33)	9 (60)	—	—	—	—	71	4.6
	Shoulder	—	—	—	—	—	8 (53)	7 (46)	—	—	—	—	67	4.4
	Arms	—	—	—	—	—	9 (60)	6 (40)	—	—	—	—	66	4.4
	Legs	—	—	—	10(66)	5 (33)	—	—	—	—	—	—	35	2.3
	Total												**297**	**19.6**
II.	Back	—	—	—	—	7(46)	5 (33)	3 (20)	—	—	—	—	56	3.7
	Neck	—	—	—	—	—	7 (46)	8 (53)	—	—	—	—	68	4.5
	Shoulder	—	—	—	—	2 (13)	5 (33)	7 (46)	—	—	—	—	74	4.9
	Arms	—	—	—	—	1 (6.6)	7 (46)	7 (46)	—	—	—	—	66	4.4
	Hand	—	—	—	—	9 (60)	6 (40)	—	—	—	—	—	51	3.4
	Legs	—	—	—	—	15 (100)	—	—	—	—	—	—	45	3.0
	Total												**360**	**24.0**
III.	Back	—	—	—	—	5 (33)	8 (53)	2 (13)	—	—	—	—	57	3.8
	Neck	—	—	—	—	2 (13)	5 (33)	7 (40)	—	—	—	—	61	4.0
	Shoulder	—	—	—	—	8 (53)	5 (33)	2 (13)	—	—	—	—	75	5.2

(Contd.)

Table 4.10.6 : (Contd.)

	Arms	—	—	—	—	7 (46)	7 (46)	1 (6.6)	—	—	—	—	54	3.6
	Hand	—	—	—	—	10 (66.6)	5 (33)	—	—	—	—	—	66	4.4
	Legs	—	—	—	—	4 (36)	9 (60)	2 (13)	—	—	—	—	58	3.8
	Total												**371**	**24.7**
IV.	Back	—	—	—	—	1 (6.6)	5 (33)	9 (60)	—	—	—	—	68	4.5
	Neck	—	—	—	—	3 (20)	0(66)	2 (13)	—	—	—	—	59	3.9
	Shoulder	—	—	—	—	2 (13)	18 (53)	5 (33)	—	—	—	—	63	4.2
	Arms	—	—	—	—	1 (6.6)	9 (60)	5 (33)	—	—	—	—	64	4.2
	Hand	—	—	—	—	15 (100)	—	—	—	—	—	—	45	3.0
	Legs	—	—	—	9 (60)	6 (40)	—	—	—	—	—	—	36	2.0
	Total												**329**	**22.1**
	Grand Total												**1333**	**22.6**

Note : Figures in parentheses represent percentage.

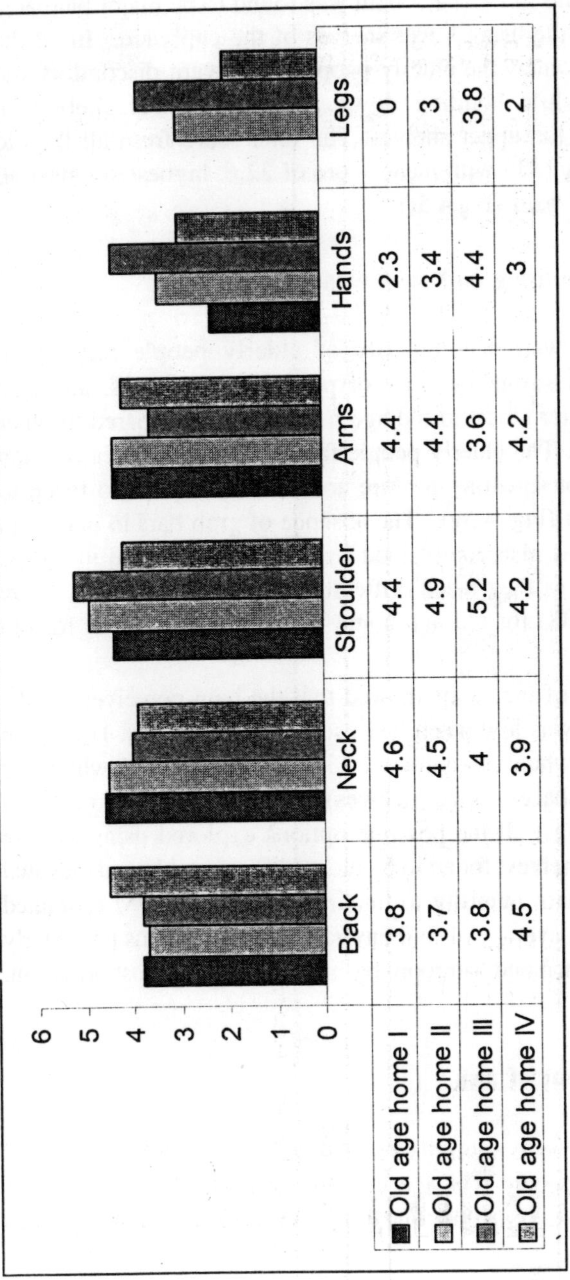

	Back	Neck	Shoulder	Arms	Hands	Legs
Old age home I	3.8	4.6	4.4	4.4	2.3	0
Old age home II	3.7	4.5	4.9	4.4	3.4	3
Old age home III	3.8	4	5.2	3.6	4.4	3.8
Old age home IV	4.5	3.9	4.2	4.2	3	2

Fig. 4.5(f) : Distribution of elderly people by their scores on perceived pain while using storage spaces (lower shelves)

IIIrd shoulder and in IVth back pain was found to be major pain area of the body while using lower shelves of the cupboard. In all the four old age homes, the elderly people felt severe discomfort and strong pain while bending down for using lower shelves in comparison to the upper shelves. The total score from all the old age homes was 1333 with mean score of 22.6, highest amongst all the features of pain (Fig.4.5f).

4.10.1.7 While using bath with water closet

The result showed that majority of elderly people felt severe discomfort and strong intensity of pain in back, knees, and legs while using water closet in old age home IInd as compared to other old age homes. The elderly people faced difficulties because they needed to adopt squatting posture and also used steps to reach to the water closet (Fig. 4.5g). The absence of grab bars in bath with water closet was also complained by the elderly people in all the old age homes as they faced difficulties of standing up. The score was highest (138) for old age home IVth and lowest (100) for old age home Ist.

A scrutiny of the study showed that the pain perceived by the elderly people was low when bed used instead of chair. High pain was perceived when ventilator used instead of window, while low when storage spaces (upper shelves) were used instead of lower shelves. Amongst all the possible options explored using storage spaces (lower shelves) found to be most uncomfortable and elevated the perceived pain intensity to maximum. *Gitlin (1998)* estimated that there are an average of four environmental problems per elderly household, kitchen and bathroom problems being the most prevalent.

4.11 Assessment of risk

Assessing risk means carefully examining what, in our work, could cause harm to people. The scoring was done on the sole basis of responses of elderly people and their perception of risk for each

Table 4.10.7 : Distribution of elderly people by their intensity of pain while using bath with water closet

Old age home	Body parts	Rating of Intensity of Pain											Total score	Mean score
		0 Nothing at all	0.5 Extremely weak	1 Very weak	2 Weak	3 Moderate	4 Some-what strong	5 Strong	6 In between strong and v. strong	7 Very strong	8 & 9 In between strong & extremely strong	10 Extremely strong		
I.	Legs	—	—	—	—	10 (66)	5 (33)	—	—	—	—	—	50	3.3
	Knee	—	—	—	—	10 (66)	5 (33)	—	—	—	—	—	50	3.3
	Total												**100**	**6.6**
II.	Legs	—	—	—	—	2 (13)	5 (33)	8 (53)	—	—	—	—	66	4.4
	Knee	—	—	—	—	3 (20)	6 (40)	6 (40)	—	—	—	—	63	4.2
	Total												**129**	**13.0**
III.	Legs	—	—	—	—	7 (40)	5 (33)	3 (20)	—	—	—	—	56	3.7
	Knee	—	—	—	—	4 (36)	10 (66)	1(6.6)	—	—	—	—	57	3.8
	Total												**113**	**7.5**
IV.	Hand	—	—	—	—	15 (100)	—	—	—	—	—	—	45	3.0
	Legs	—	—	—	—	12 (80)	3 (20)	—	—	—	—	—	48	3.2
	Knee	—	—	—	—	15 (100)	—	—	—	—	—	—	45	3.0
	Total												**138**	**9.2**
	Grand Total												**480**	**8.0**

Note : Figures in parentheses represent percentage.

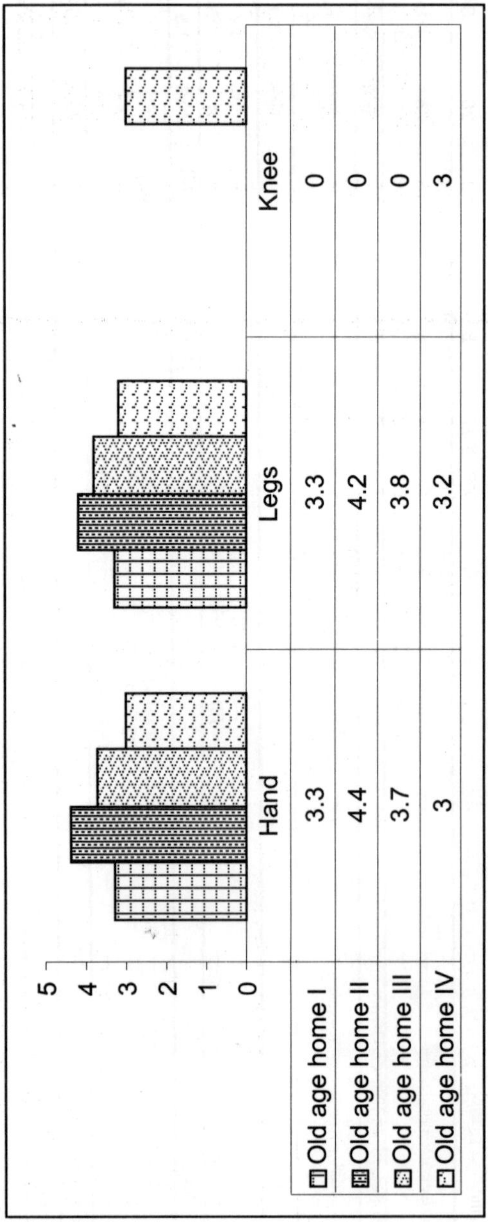

Fig. 4.5(g) : Distribution of elderly people by their scores on perceived pain while using bath with water closet

Table 4.11.1 : Assessment of risk perceived by the elderly people

Description	*Old age homes*								*Total score F = 60*	
	I		*II*		*III*		*IV*			
	M = 9	*F = 6*	*M = 8*	*F = 7*	*M = 7*	*F = 8*	*M = 2*	*F = 13*	*M = 26*	*F= 34*
Inadequate entrance door	—	—	—	—	—	—	—	—	—	—
Insufficient lighting in bath, stairs, corridors	9 (100)	6 (100)	8 (100)	7 (100)	7 (100)	8 (100)	2 (100)	13 (100)	26	34
Absence of night light	9 (100)	6 (100)	8 (100)	7 (100)	6 (85)	8 (100)	2 (100)	13 (100)	25	34
Absence of generator facility	—	—	1 (12.5)	7 (100)	7 (100)	8 (1.00)	2 (100)	13 (100)	10	28
Absence of bell for emergency	9 (100)	6 (100)	8 (100)	7 (100)	7 (100)	8 (100)	2 (100)	13 (100)	26	24
Unreachable height. of door, window latches	—	6 (100)	1 (1.25)	7 (100)	—	8 (100)	—	13 (100)	1	34
Unreachable height. of ventilator	3 (33)	6 (100)	2 (25)	7 (100)	3 (42)	8 (100)	—	13 (100)	8	34
Inaccessibility of bath with water closet	—	—	—	—	—	—	—	11 (84)	—	11
Squatting toilet (lower seat)	—	—	7 (87)	7 (100)	—	—	—	—	7	7
Absence of grab bars in bathroom	9 (100)	6 (100)	7 (87)	7 (100)	5 (71)	8 (100)	2 (100)	13 (100)	23	34
Slippery and uneven flooring	2 (22)	3 (50)	6 (75)	7 (100)	4 (57)	8 (100)	2 (100)	11 (84)	14	29

(Contd.)

Table 4.11.1 : (Contd.)

Inaccessibility of dinning spaces	5 (55)	5 (83)	—	—	1 (14)	7 (87)	2 (100)	13 100)	8	25
Height of storage spaces in room	—	3 (50)	3 (37)	7 (100)	—	—	2 (92)	12 (100)	5	22
Height of storage spaces in bathroom	1 (11)	2 (33)	3 (37)	7 (100)	—	—	—	—	4	9
Absence of handrails at the top and bottom of stairs	—	—	—	—	—	—	—	—	—	—
Obstructed pathways	—	—	—	—	—	—	—	—	—	—
Extreme low/high height of bed	2 (22)	4 (66)	2 (25)	3 (42)	5 (71)	6 (75)	1 (50)	12 (92)	10	25
Raised structure/equipment	—	6 (100)	8 (100)	7 (100)	—	2 (25)	—	—	8	15
Absence of rubber mats in wet area	9 (100)	6 (100)	8 (100)	7 (100)	6 (85)	8 (100)	2 (100)	13 (100)	25	34
Bolts/nuts/sharp edges of furniture	2 (22)	2 (33)	1 (12.5)	2 (28)	1 (14)	2 (25)	—	1 (7.6)	4	7
Improper drainage	7 (77)	6 (100)	5 (62.5)	4 (57)	5 (71)	4 (50)	2 (100)	13 (100)	19	27
Improper clothing/footwear	—	3 (50)	2 (25)	7 (100)	1 (14)	2 (25)	1 (50)	2 (15.3)	4	14
Steps too high	3 (33)	5 (83)	3 (33)	5 (71)	1 (14)	2 (25)	1 (50)	10 (76)	8	22

Note : Figures in parentheses represent percentage.

M = Male, F= Female

statement. For the quantification of qualitative data numerical values was assigned to the response, *i.e.* for Yes response received was denominated one as score which was interpreted as risk involved, whereas, for the response No, no numerical value, *i.e.* zero was interpreted as no risk, whereas higher the 1 scores greater the risk. The scores earned by elderly people were categorized in three ranges, *i.e.* 1-7 scores, 8-15 scores, and 16-23 scores and interpreted as low, moderate and high risk, respectively. The findings are summarized in Table 4.11.1, showed that the entrance door of all the old age homes were out of risk as perceived by both the sexes. But insufficient lightings in bathroom, stairs, corridors, absence of number of luminaries for light during night hours, lack of generator facilities during light cut off, lack of bell at the time of emergencies, absence of grab bars in bathroom, slippery flooring, improper drainage during rainy season, absence of rubber mats in likely to be wet area, *i.e.* bathroom were perceived risk by both the sexes. The elderly females perceived higher risk for features like — unreachable height of ventilator, latches of door, window, and storage spaces. It is inferred that the females of the lot were found to be more prone to risk for most of the parameters as compared to the elderly males (Fig. 4.6).

4.11.1 Extent of risk perceived by the elderly people

On the basis of data obtained for risk assessment the elderly people had been categorized into low, moderate and high range of risk. The findings are summarized in Table 4.11.2 showed that 73 per cent of elderly males and 35 per cent of females were found to be in moderate risk range, while 61.7 per cent of females and 11.5 per cent of males in high risk zones. The elderly people belonging to the old age home IInd though were found to be high risk prone yet they affirmed to be satisfied with the available facilities. They seemed to have adjusted with the existing situation. In old age homes Ist and IVth the elderly people were satisfied but they perceived higher risk.

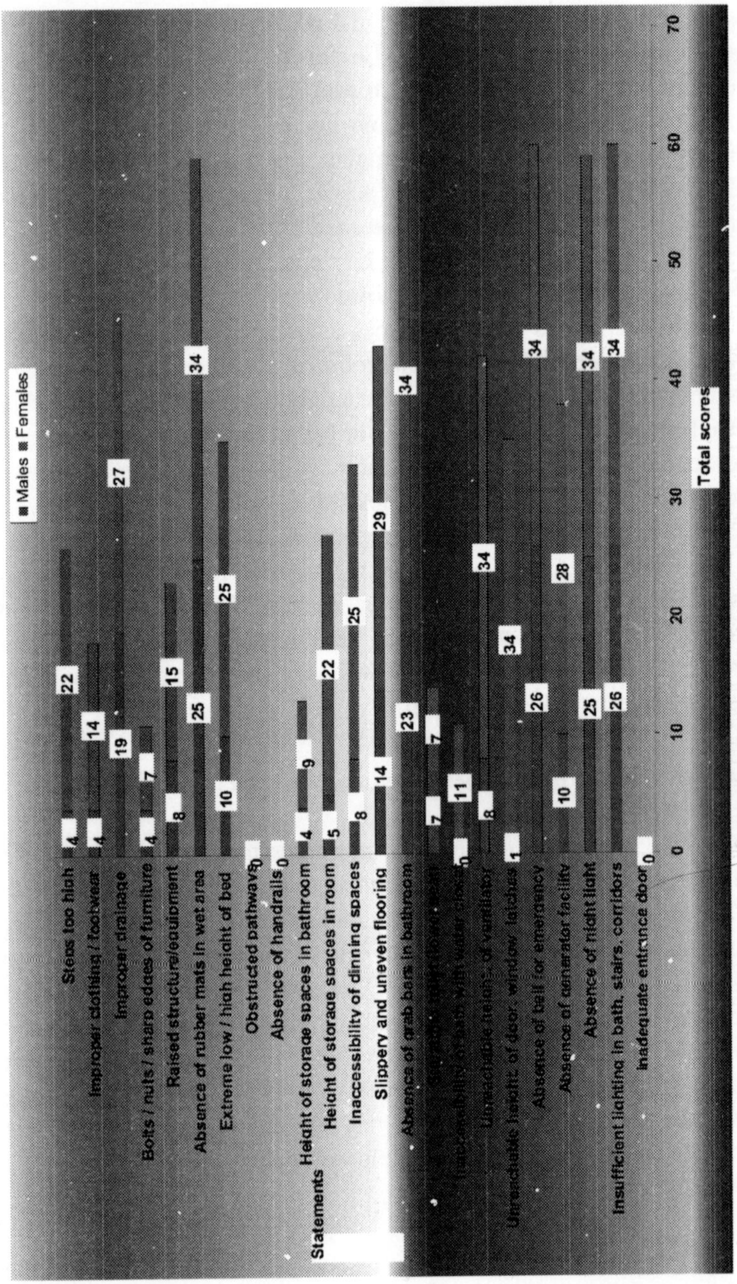

Fig. 4.6 : Assessment of perceived Risk

Table 4.11.2 : Distribution of elderly people by their scores on risk perceived

Score	*Range*	*Old age home*								*Total F = 60*	
		I		*II*		*III*		*IV*			
		M = 9	*F = 6*	*M = 8*	*F = 7*	*M = 7*	*F=8*	*M =2*	*F = 13*	*M = 26*	*F = 34*
1–7	Low	1(11)	–	–	–	2(28.5)	1(12.5)	–	–	3(11.5)	1(2.9)
8–15	Moderate	8(88)	1(16)	5(62.5)	–	5(71)	7(87.5)	2(100)	4(30)	20 (76.9)	12(35.2)
16–23	High	–	5(83)	3(37.5)	7(100)	–	–	–	9(69)	3(11.5)	21(61.7)

Note : Figures in parentheses represent percentage.

M = Male, F = Female

4.12 Ergonomic assessment of old age homes

It deals with the assessment of old age homes on the basis of ergonomic principles. A large number of factors play a role in ergonomics; these include body posture, work place designing and environmental factors (noise, temperature, humidity illumination, and climate). The checklist was made to assess the old age homes ergonomically. The findings are summarized in Table 4.12 showed that the habitat rooms were constructed according to the National Building Codes, whereas the separate baths were not. The ceilings and walls were sound and free from hazardous defects like damaged, large holes etc. in all the four old age homes. But in old age homes IInd and IIIrd, the roof did not permit effectual drainage of rain waters, which caused dampness in the wall. The dampness of wall affected the health of the elderly people causing respiratory problems. The interior finishes of wall were in good condition. The doors and doorways had clear openings when opened and were operable by single efforts. The material used for door was not resistant against moisture. This again adversely affected the health of elderly people. The electrical wiring was in proper condition so as to ensure safety. Absences of safety architectural symbols like main fuse board, switch outlets, fan and regulator etc. in electric installation were the major shortcoming that caused inconvenience and hazards safety. The floorings of the habitat rooms were not specific, smooth and of grease proof material (skid resistance tiles) particularly in bath with water closets to avoid any accidental slips/falls which were the major drawback. The floor of the bath with water closets was not sloped towards the drain with a suitable grade and also not uniform but was raised to make the difference between bath and water closet in old age home IInd. The corner between the wall were not rounded which make elderly people prone to accident. Major concerns especially for elderly females were the height of windows, ventilators, storage spaces, and also the furniture. The fittings and fixtures like that of bath and electric were not suitable and not made keeping into consideration the body dimension of elderly people. The environmental parameters specially the illuminations level was not friendly to the elderly people and caused accidental injuries. Good

Table 4.12 : Ergonomic assessment of old age homes

Description	*Old age home*			
	I Score	*II Score*	*III Score*	*IV Score*
1. Habitat room	1	1	1	1
2. Bath with water closet	1	1	1	1
3. Ceiling free from hazardous defects	1	1	1	1
4. Walls free from hazardous defects	1	1	1	1
5. Permanent passageways marked	0	0	0	0
6. Roof permit effectual drainage of rain water	1	0	0	1
7. Door and doorways have clear opening	1	1	1	1
8. Enough number of door, window, ventilator	1	1	1	1
9. Cross ventilation	0	1	1	1
10. Proper electric wiring	1	1	1	0
11. Height of electric switches	0	0	1	0
12. Safety architectural symbols on electric installation	0	0	0	1
13. Floor designed evenly	1	1	1	1
14. Stairways safe and no slippery	1	1	1	0
15. Accessibility of bath with water closet	1	1	1	0
16. Minimum fresh air requirement for toilet exhaust	0	0	0	0
17. Flooring of bathroom of grease proof	0	0	0	0
18. Bathroom floor sloped suitably towards the drains	0	0	0	0
19. Uniform/elevated /depressed floors	1	0	1	0
20. Corner between the walls rounded	0	0	0	0
21. Chair designed ergonomically	0	0	0	0
22. Bed designed ergonomically	0	0	0	0
23. Table designed ergonomically	0	0	0	1
24. Fittings and fixtures designed ergonomically	0	0	1	0
25. Height of window, ventilator suitably designed	0	0	0	0
26. Height of the storage spaces	0	0	0	0
27. Environmental parameters	0	0	0	0
Total Score	**12**	**11**	**14**	**11**

lighting not only assists with reading and safety, it helps prevent depression for those confined indoors. If cost is a concern, the improved, inexpensive fluorescent light bulbs may be satisfactory. The total score obtained was highest (14) for old age home IIIrd and lowest (11) for old age home I[st].

On the grounds of findings, it was inferred that the old age home IIIrd was superior to other old age homes for the elderly people but still require to be building ergonomically.

4.13 Extent of satisfaction derived by the elderly people

Extent of satisfaction derived by elderly people from old age home was one of the dependent variable in the present investigation. In order to get meaningful result, a descriptive satisfaction scale was constructed to assess the extent of satisfaction derived. Responses on each item in the scale were scored for highly satisfied (3), satisfied (2) and not satisfied (1). From the individual item score total and mean score of on each item of each elderly people was computed. This study models the impact of architectural and psycho-social variables on house satisfaction of elderly.

The findings related to frequency distribution, total and mean score on each statement of the scale are presented in Table 4.13.1.

4.13.1 Statement of satisfaction

4.13.1.1 Size of room

Data in Table 4.13 showed that 33 and 6 per cent of elderly people were highly satisfied in old age homes IIIrd, and IVth respectively, while 100 per cent satisfied with size of the room in old age homes Ist, and IInd, and only 13 per cent not satisfied in old age homes IVth. Total score obtained was 124 with mean score of 2.0.

4.13.1.2 Size of door

In old age homes Ist, IInd, IIIrd, and IVth, 6, 13, 100, and 6 per cent elderly people, respectively, were highly satisfied with door sizes, while 100, 66, and 50 per cent were satisfied. None of them was dissatisfied with sizes of the doors in all the four old age homes. Total score obtained was 109 with mean score of 1.8.

Table 4.13.1 : Extent of satisfaction dervived by the elderly people about old age homes

Statement	Old Age Home				Old Age Home				Old Age Home				Total score	Mean score
	I F=15	II F=15	II F=15	IV F=15	I F=15	II F=15	III F=15	IV F=15	I F=15	II F=15	III F=15	Iv F=15		
	Highly Satisfied = 3				*Satisfied = 2*				*Note Satisfied = 1*					
1. Size of room	—	—	5 (33)	1 (6)	15 (100)	15 (100)	10 (66)	12 (80)	—	—	—	2 (13)	124	20
2. Size of door	1 (6)	2 (13)	15 (100)	1 (6)	14 (93)	13 (86)	—	14 (93)	—	—	—	—	109	1.8
3. Size of window	—	5 (33)	10 (66)	—	14 (93)	10 (66)	5 (33)	15 (100)	1 (6)	—	—	—	134	2.2
4. Size of ventilator	—	—	—	—	13 (86)	13 (86)	15 (100)	15 (100)	2 (13)	2 (13)	—	—	116	1.9
5. Height of cupboard	3 (20)	—	—	—	12 (80)	5 (33)	6 (40)	4 (26)	—	10 (66)	9 (60)	11 (73)	93	1.5
6. Height of shelves	—	—	—	—	3 (20)	11 (73)	10 (66)	7 (46)	12 (80)	4 (26)	5 (33)	8 (53)	91	1.5
7. Working with door	—	3 (20)	1 (6)	—	13 (86)	12 (80)	14 (93)	11 (73)	2 (13)	—	—	4 (26)	118	1.9
8. Working with window	—	—	—	—	11 (73)	11 (73)	9 (60)	13 (86)	4 (26)	4 (26)	6 (40)	2 (13)	104	1.7
9. Working with ventilator	—	—	—	—	2 (13)	3 (20)	4 (26)	1 (6)	13 (86)	12 (80)	11 (73)	14 (93)	70	1.1
10. Direction of door	—	5 (33)	5 (33)	3 (20)	11 (73)	10 (66)	10 (66)	12 (80)	4 (26)	—	—	—	129	2.1
11. Direction of window	—	—	5 (33)	2 (13)	10 (66)	2 (13)	10 (66)	13 (86)	5 (33)	13 (86)	—	—	74	1.2

(Contd.)

Table 4.13.1 : (Contd.)

12. Direction of ventilator	—	3 (20)	—	1 (6)	5 (33)	12 (80)	15 (100)	14 (93)	10 (66)	—	—	—	102	1.7
13. Handles of door and window	—	6 (40)	3 (20)	1 (6)	11 (73)	9 (60)	12 (80)	14 (93)	4 (26)	—	—	—	126	2.1
14. Condition of walls, ceiling	2 (13)	4 (26)	—	—	13 (86)	11 (73)	5 (33)	11 (73)	—	—	10 (66)	4 (26)	112	1.8
15. Condition of floor	—	—	—	—	9 (60)	3 (20)	8 (53)	9 (60)	6 (40)	12 (80)	7 (46)	6 (40)	89	1.4
16. Number of furniture	—	5 (33)	10 (66)	2 (13)	9 (60)	8 (53)	5 (33)	9 (60)	6 (40)	2 (13)	—	4 (26)	125	2.0
17. Size and shape of chair	—	6 (40)	—	—	6 (40)	9 (60)	11 (73)	10 (66)	9 (60)	—	4 (26)	5 (33)	88	1.4
18. Size and shape of table	—	5 (33)	5 (33)	1 (6)	11 (73)	5 (33)	10 (66)	14 (93)	4 (26)	5 (33)	—	—	120	2.0
19. Size and shape of bed	—	5 (33)	—	—	7 (46)	10 (66)	7 (46)	6 (40)	8 (53)	—	8 (53)	9 (60)	100	1.6
20. Mattress and cushion	—	7 (46)	2 (13)	—	5 (33)	8 (53)	9 (60)	8 (53)	10 (66)	—	4 (26)	7 (46)	108	1.8
21. Size of bathroom and W.C.	—	8 (53)	5 (33)	—	4 (26)	7 (46)	6 (40)	11 (73)	11 (73)	—	4 (26)	4 (26)	104	1.7
22. Facilities in bathroom	—	—	—	—	5 (33)	6 (40)	10 (66)	2 (13)	10 (66)	9 (60)	5 (33)	13 (86)	83	1.3
23. Internal temperature	—	—	—	—	7 (46)	4 (26)	15 (100)	15 (100)	8 (53)	11 (73)	—	—	101	1.6
24. Sanitary conditions (waterlogging)	—	—	5 (33)	—	4 (26)	8 (53)	10 (66)	15 (100)	11 (73)	7 (46)	—	—	96	1.6
25. Noise level	—	2 (13)	7 (46)	1 (6.6)	10 (66)	6 (40)	8 (53)	14 (93)	5 (33.3)	7 (46)	—	—	118	1.9
26. Daily activities	5 (33)	13 (86)	—	—	10 (66)	2 (13)	15 (100)	15 (100)	—	—	—	—	138	2.3

(Contd.)

Table 4.13.1 : (Contd.)

27. Number and type of luminaries	3 (20)	4 (26)	—	—	3 (20)	9 (60)	5 (33)	6 (40)	9 (60)	2 (13)	10 (66)	9 (60)	82	1.3
28. Natural lighting	1 (6.6)	8 (53)	—	1 (6.6)	10 (66)	7 (46)	7 (46)	9 (60)	4 (26)	—	8 (53)	4 (26)	112	1.8
29. Place of electric switches	—	—	3 (20)	—	3 (20)	11 (73)	12 (80)	6 (40)	12 (80)	4 (26)	—	9 (60)	98	1.6
30. Recreational facilities	—	5 (33)	—	—	—	5 (33)	15 (100)	10 (66)	15 (100)	5 (33)	—	5 (33)	100	1.6
31. Medical facilities	—	4 (26)	—	—	7 (46)	9 (60)	15 (100)	5 (33)	8 (53)	2 (13)	—	10 (66)	104	1.7

Note : Figures in parentheses represent percentage.

4.13.1.3 Size of window

33 and 66 per cent of elderly people were highly satisfied in old age homes IInd and IIIrd, respectively, while 93, 66, 33 and 100 per cent were satisfied with window size in old age homes Ist, IInd, IIIrd and IVth, respectively, and only 6 per cent were not satisfied in old age homes Ist. Total score obtained was 134 with mean score of 2.2 which was highest compared to other architectural features of house in the satisfaction scale.

4.13.1.4 Size of ventilator

Equal percentage, *i.e.* 86 per cent, and 100 per cent of elderly people were satisfied in old age homes Ist, IInd, IIIrd and IVth, while none of them highly satisfied and remaining were not satisfied with respect to the size of ventilator. Total score obtained was 116 with mean score of 1.9.

4.13.1.5 Height of cupboard

20 per cent of elderly people in old age homes Ist were highly satisfied, while 93, 66, 33 and 100 per cent were satisfied in old age homes Ist, IInd, IIIrd and IVth, respectively and remaining were not satisfied, with respect to the height of cupboard, as it was too high and most of them found it difficult to reach. Total score obtained was 93 with mean score of 1.5.

4.13.1.6 Height of shelves

It was found that 20, 73, 66 and 46 per cent of elderly people were satisfied, while remaining were not satisfied in old age homes Ist, IInd, IIIrd and IVth, respectively. None of them was highly satisfied in all the four old age homes. Elderly people experience particular problems with the height and location of cupboards. Total score obtained was 91 with mean score of 1.5.

4.13.1.7 Working with door

20 per cent and 6 per cent in old age homes Ist, IInd, were highly

satisfied respectively, while majority of them 86 per cent, 80 per cent, 93 per cent, and 17 per cent were satisfied in old age homes Ist, IInd, IIIrd and IVth, respectively, and only 13 per cent and 26 per cent were not satisfied in old age homes Ist and IVth, respectively. Total score obtained was 115 with mean score of 1.9.

4.13.1.8 Working with window

70 per cent each in old age homes Ist and IInd and 60 and 86 per cent in old age home IIIrd and IVth were satisfied with windows, remaining were not satisfied. None of them was highly satisfied. Total score obtained was 104 with mean score of 1.7.

4.13.1.9 Working with ventilator

In old age homes Ist, IInd, IIIrd and IVth, 13, 20, 26 and 6 per cent of elderly people were satisfied with respect to working with ventilator, respectively, while remaining not satisfied. None of them was highly satisfied in all the four old age homes. This feature had scored lowest total score 69 with mean score of 1.1 compared to other features, which denoted least satisfaction among elderly people.

4.13.1.10 Direction of door

Equal percentage (33 %) of elderly people in old age homes IInd, and IIIrd and 20 per cent in old age home IVth were highly satisfied with direction of the door, while majority 73, 80 per cent in old age homes Ist, IVth, and 66 per cent each in old age homes IInd and IIIrd were satisfied, respectively, while only (26 %) in old age home Ist were not satisfied. Total score obtained was 129 with mean score of 2.1.

4.13.1.11 Direction of window

33 and 13 per cent of elderly people in old age homes IIIrd and IVth, respectively, were highly satisfied while, 60, 13, 66 and 86 per cent were satisfied in old age homes Ist, IInd, IIIrd and IVth,

respectively and only 33 and 86 per cent were not satisfied in old age homes Ist and IInd, respectively with respect to the direction of window. Total score obtained was 74 with mean score of 1.2.

4.13.1.12 Direction of ventilator

20 and 6 per cent of elderly people were highly satisfied in old age homes IInd, and IVth, respectively, while 33, 80, 100 and 93 per cent were satisfied in old age homes Ist, IInd, IIIrd and IVth, respectively and 66 per cent were not satisfied.

4.13.1.13 Handles of door and window

73, 60, 80 and 93 per cent of elderly people were satisfied in old age homes Ist, IInd, IIIrd and IVth, respectively, while 40, 20 and 6 per cent in old age homes IInd, IIIrd and IVth, respectively were highly satisfied and 26 per cent were not satisfied in old age home Ist. Total score obtained was 126 with mean score of 2.1.

4.13.1.14 Condition of wall and ceiling

86, 73, 33 and 73 per cent of elderly people were satisfied in old age homes Ist, IInd, IIIrd and IVth, respectively, while 13 and 26 per cent were highly satisfied in old age homes Ist and IInd, respectively and 66 and 26 per cent were not satisfied with the condition of wall and ceiling in old age homes IIIrd and IVth, respectively due to the damp and defects. Total score obtained was 112 with mean score of 1.8.

4.13.1.15 Condition of floor

In old age homes Ist, IInd, IIIrd and IVth, it was found that 60, 20, 53 and 60 per cent of elderly people were satisfied, respectively, while remaining were not satisfied with conditions of floor. None of them was highly satisfied in all the four old age homes. Total score obtained was 89 with mean score of 1.4.

4.13.1.16 Number of furniture

It was found that 33, 66, and 13 per cent of elderly people were highly satisfied in old age homes Ist, IInd, IIIrd and IVth, respectively, while 60, 53, 33, 60 and 40 per cent were satisfied in old age homes Ist, II^{n}, IIIrd and IVth, respectively remaining were not satisfied. Total score obtained was 125 with mean score of 2.0.

4.13.1.17 Size and shape of the chair

In old age home IInd, 33 per cent of elderly people were highly satisfied, while 40, 60, 33 and 66 per cent were satisfied in old age homes Ist, IInd, IIIrd and IVth, respectively and 60, 26 and 33 per cent were not satisfied in old age homes Ist, IIIrd and IVth, respectively with respect to size and shape of the chair. Total score obtained was 88 with mean score of 1.4.

4.13.1.18 Size and shape of table

33 per cent each in old age home IInd and, IIIrd and 6 per cent in IVth, respectively were highly satisfied, while 73, 33, 66 and 93 per cent were satisfied in old age homes Ist, IInd, IIIrd and IVth, respectively and 26 per cent and 33 per cent were not satisfied with respect to size and shape of table in old age homes Ist and IInd, respectively. Total score obtained was 120 with mean score of 2.0.

4.13.1.19 Size and shape of bed

It was found that only 33 per cent of elderly people were highly satisfied in old age home IInd, while 46, 66, 46 and 40 per cent were satisfied in old age homes Ist, IInd, IIIrd and IVth, respectively, In old age homes Ist and IIIrd each 53 per cent, while in IVth 60 per cent were not satisfied, respectively. Total score obtained was 100 with mean score of 1.6.

4.13.1.20 Mattress and cushion

It was found that only 46 and 13 per cent of elderly people were highly satisfied in old age homes IInd and IIIrd, respectively, while

33, 53, 60 and 53 per cent were satisfied in old age homes Ist, IInd, IIIrd and IVth, respectively, and 66, 26, and 46 per cent were not satisfied in old age homes Ist, IIIrd and IVth, respectively. Total score obtained was 108 with mean score of 1.8.

4.13.1.21 Size of bathroom with water closet

It was found that only 53 and 33 per cent of elderly people were highly satisfied in old age homes IInd and IIIrd, while 26, 46 and each 40 per cent were satisfied in old age homes Ist, IInd, IIIrd and IV^{th,} respectively, while 73, each 26 per cent were not satisfied in old homes Ist, IIIrd and IVth, respectively. Total score obtained was 109 with mean score of 1.8.

4.13.1.22 Facilities in bathroom with water closet

The bathroom attraeted most comment, it was found that 66, 60, 33, and 86 per cent were not satisfied, while only 33, 40, 66, 13 per cent, elderly people were satisfied in old age homes Ist, IInd, IIIrd and IVth, respectively. None of them was highly satisfied with condition of facilities in bathrooms with water closets in all the old age homes. All improvements suggested by older people related to toilet facilities and bathrooms. Total score obtained was 83 with mean score of 1.3.

4.13.1.23 Internal temperature

It was found that 46, 26 and each 100 per cent of elderly people were satisfied in old age homes Ist, IInd IIIrd and IVth, respectively, while 53 and 73 per cent were not satisfied in old age homes Ist and IInd, respectively. None of them were highly satisfied with respect to internal temperature in all the old age homes as no facilities were made during temperature extremes. Total score obtained was 101 with mean score of 1.6.

4.13.1.24 Sanitary condition

It was found that only 33 per cent of elderly people in old home IIIrd were highly satisfied, while 26, 53, 66 and 100 per cent were satisfied in old age homes Ist, IInd, IIIrd, and IVth, respectively, while 33 and 73 per cent were not satisfied with the sanitary condition in Ist and IInd old age homes, as there was waterlogging problem during rainy season. Total score obtained was 96 with mean score of 1.6.

4.13.1.25 Noise level

It was found that 13, 46 and 6.6 per cent elderly people were highly satisfied in old age homes IInd, IIIrd and IVth, respectively, while 66, 40, 53 and 93 per cent were satisfied, remaining 46 and 33 per cent were not satisfied in old age homes Ist and IInd, respectively as there was always some religious and school functions carried out within the vicinity of the premises. Total score obtained was 118 with mean score of 1.9.

4.13.1.26 Daily activities

In old age homes Ist, IInd, 33 and 86 per cent of elderly people were highly satisfied, respectively, while 66.6, 13 and each 100 per cent were satisfied in Ist, IInd, IIIrd and IVth old age homes, respectively. None of them was dissatisfied with respect to daily scheduled activities. Total score obtained was highest (138 with mean score of 2.3) compared to other features which denoted that the elderly people were highly satisfied.

4.13.1.27 Number and type of luminaries

In old age homes Ist, IInd, 20 and 26 per cent of elderly people were highly satisfied, respectively, while 20, 66, 33, and 46 per cent in old age homes Ist, IInd, IIIrd and IVth, respectively satisfied and 73, 13, 66 and 60 per cent were not satisfied in old age homes Ist, IInd, IIIrd and IVth, respectively with respect to number and type of luminaries. Total score obtained was 101 with mean score of 1.6.

4.13.1.28 Natural lighting

In old age homes Ist, IIIrd and IVth, 66, 53 and 6.6 per cent of elderly people were highly satisfied, while 66.6, 46 per cent each and 60 per cent satisfied in old age homes Ist, IInd, IIIrd and IVth, respectively and 26, 53 per cent each were not satisfied in old age homes Ist, IIIrd and IVth, respectively. Total score obtained was 112 with mean score of 1.8.

4.13.1.29 Placement of electric switches

In old age home IIIrd, only 20 per cent elderly people were highly satisfied, while 20, 73, 80 and 40 per cent were satisfied in old age homes Ist, IInd, III^{rd,} and IVth, respectively, while 80 and 60 per cent were not satisfied in old age homes Ist and IVth, respectively. Total score obtained was 98 with mean score of 1.6.

4.13.1.30 Recreational facilities

In old age home IInd only 33 per cent of elderly people were highly satisfied, while 33, 100, and 66 per cent were satisfied in IInd, IIIrd and IVth old age homes, respectively, and 100 per cent, each 33 per cent were not satisfied in old age homes Ist, IInd and IVth, respectively. Total score obtained was 100 with mean score of 1.6.

4.13.1.31 Medical facilities

In old age home IInd only 26 per cent of elderly people were highly satisfied, while 33, 100, 66 per cent were satisfied in IInd, IIIrd and IVth old age homes, respectively, and 53, 13 and 66 per cent were not satisfied in old age homes Ist, IInd and IVth, respectively. Total score obtained was 104 with mean score of 1.7.

4.13.2 Extent of satisfaction derived

The scores earned by elderly people was categorized in three ranges, *i.e.* 40-50 scores, 51-61 scores, and 62-72 scores and interpreted as low, moderate and high level of satisfaction, respectively.

Overall working with ventilator had earned the lowest total

score of 69 with mean score of 1.1, because the elderly females faced difficulties due to increased height, while daily scheduled activities of the old age home had earned the highest total score 138 with mean score of 2.3. Table 4.13.2 illustrated that 65 per cent of elderly people were found to be in moderate range, while 23.3 per cent in high range of satisfaction level. The reasons may be that the old age homes were provided an ambience that will ward off depression and give them a feeling of purpose and belongingness as the authority immerses them in wholesome and worthwhile activities that foster a sense of community among their peers. There is a widespread belief that as older people progress through retirement, their satisfaction with their house increases.

Table 4.13.2 : Distribution of elderly people by their scores on extent of satisfaction derived

Score	*Range*	*Old age home*				
		*I^st^*	*IInd*	*IIIrd*	*IVth*	*Total*
40-50	Low	7(46.6)	—	—	—	7(11.6)
51-61	Moderate	8(53)	8(53)	8(53)	15(100)	39(65)
62-72	High	—	7(46.6)	7(46.6)	—	14(23.3)

Note : Figures in parentheses represent percentage.

4.14 Testing of Hypotheses

In order to test the hypotheses statistically, the null hypotheses as presented below were formulated for each hypothesis.

4.14.$H_0$1 : The age and sex of elderly people have no relationship with occurrence of accidents.

Table 4.14.1 : Chi-square values for testing the relationship between occurrence of accidents with age and sex of the elderly people

Old age home	*Chi-square value*	*Conclusion*
I^st^	0.0001	NS characters are independent
IInd	1.7013	NS characters are independent
IIIrd	4.1201*	Significant characters are dependent
IVth	3.9310*	Significant characters are dependent

NS: Not significant

It is evident from the chi-square calculated value that in old age homes Ist and IInd, there was no significance relation between the occurrence of accidents and the age and the sex of elderly people. It is concluded that the occurrence of accidents and the age and sex of the elderly people are independent factors, whereas in

Table 4.14.2 : Chi-square values for testing the relationship between anthropometrical measurements of elderly people and dimensions of (a) storage areas and (b) furniture

Description	*Chi-square Value*	*Conclusion*
1. Relationship between height of elderly people and		
a. Length of window	1.621	NS characters are independent
b. Length of door	0.101	NS characters are independent
2. Relationship between vertical reach and		
a. Height of cupboard	0.086	NS characters are independent
b. Height of open rack	0.036	NS characters are independent
c. Height of built in cupboard	1.303	NS characters are independent
3. Relationship between forward reach and		
a. Depth of cupboard	0.182	NS characters are independent
b. Depth of open rack	0.013	NS characters are independent
4. Relationship between sideway reach		
a. Shelves length	0.256	NS characters are independent
5. Relationship between popliteal height and		
a. Sitting height from floor	0.000	NS characters are independent
6. Relationship between height of body above seat and		
a. Back rest height of chair	1.236	NS characters are independent
6. Relationship between elbow height and		
a. Arm rest height of chair	2.583	NS characters are independent
7. Relationship between hip breadth and		
a. Width of chair	1.701	NS characters are independent
8. Relationship between buttock to popliteal length and		
a. Seat length of chair	0.364	NS characters are independent
9. Relationship between height of bed and		
a. Length of bed	0.052	NS Characters are independent
10. Relation between hip breadth and		
a. Width of bed	3.885*	Significant characters are dependent

NS: Not significant.

IIIrd and IVth old age homes, the chi-square values showed significant results. Hence, it is inferred that the occurrence of accidents depends on the age and sex of the elderly people for IIIrd and IVth old age homes.

On the strength of these findings the above null hypotheses $H_0 1$ was partially rejected and it is inferred as the occurrence of accidents depends on the age and sex of elderly people.

4.14.$H_0$2 : Anthropometrical measurements of elderly people have no relationship with dimensions of (a) storage areas and (b) furniture.

It is evident from the chi-square calculated value that there was no significant relationship between selected anthropometric measurements of elderly people and (a) dimensions of storage spaces and (b) furniture at 5 per cent level of significance. Since the calculated value of chi-square in all the cases was less than the table value of chi-square except for one variable.

On the strength of these findings the above null hypotheses $H_0 2$ was not rejected. Hence, it is inferred that the dimensions of storage spaces and the dimensions of furniture were independent of anthropometric measurements of elderly people.

4.14.$H_0$3 : Anthropometrical measurements of elderly males have no relationship with anthropometrical measurements of elderly females.

Table 4.14.3 : Correlation coefficients between anthropometric measurements of elderly males and females

Correlation co-efficient	*r values*	*Conclusion*
The height, weight and body surface area of males and females	0.996*	Significant at 5 per cent level
Anthropometric measurements (standing) of males and females	0.99**	Highly significant at 1 per cent level
Anthropometric measurements (sitting) of males and females	0.995**	Highly significant at 1 per cent level

The height, weight and body surface area of males and females were significantly related to each other at 5 per cent level of significance. Anthropometric measurements (standing and sitting)

of males and females were highly significantly related to each other at 1 per cent level of significance.

On the strength of these findings the above null hypotheses $H_0 3$ was rejected. Hence, it is inferred that there was a significant positive correlation co-efficient between the anthropometric measurement of elderly males and anthropometric measurement of elderly females.

4.14.$H_0$4 : Dimensions of old age homes have no relationship with National Building Codes.

Table 4.14.4 : Fisher's *t*-value for testing significant difference between dimensions of old age homes and dimensions given in National Building Codes

Description	*t-value*	*Conclusion*
Dimensions of habitat room and dimensions given in National Building Codes	1.282	Not significant
Dimensions of bathroom with water closet and dimensions given in National Building Codes	1.767	Not significant
Dimension of separate water closet and dimensions given in National Building Codes	21.896*	Significant
Dimension of store room and dimensions given in National Building Codes	176.48*	Significant

It is evident from the *t*-value that there was no significant relationship between dimensions of habitat rooms and dimensions given in National Building Codes.

Hence, it is inferred that the dimensions of habitat rooms and bathroom with water closets showed equivalence with National Building Codes, whereas separate water closet, and store room showed significant relationship. On the strength of these findings the above null hypotheses $H_0 4$ was partially rejected and it is inferred as the old age homes were not constructed according to National Building Codes.

4.14.$H_0$5 : Dimensions of furniture have no relationship with Indian Standard Codes.

The finding related to null hypotheses are summarized in Table 4.14.5 the computed *t*-value showed the significant relationship for height, depth of chair, whereas for width of bed not significant. On the strength of these findings the above null hypotheses $H_0 5$ was rejected and it is inferred that the furniture provided in old age homes were not made according to the dimensions given for furniture in Indian standard codes.

Table 4.14.5 : Fisher's *t*-value for testing significant difference between dimensions of furniture and dimensions given in Indian Standard Codes

Description	*t -value*	*Conclusion*
1. Height of the chair and dimensions given in Indian Standard Codes for height of the chair	12.283*	Significant
2. Depth of chair and dimensions given in Indian Standard Codes for depth of the chair	5.794*	Significant
3. Length of bed and dimensions given in Indian Standard Codes for length of bed	5.200*	Significant
4. Width of bed and dimensions given in Indian Standard Codes for width of bed	0.899	Significant

5

Summary and Conclusion

The Indian family has traditionally provided natural social security to the old people. However, in more recent times, institutions such as old age homes are sharing the traditional role of family due to some factors like increased urbanization, migration of younger adults of the family to towns and in some cases to foreign countries and increased participation of women in paid employment outside the home. As the proportion of elderly people in the Indian population continues to increase, there are demands for increase in number of old age homes in India and consequent to these there is need to ensure safety. Until recent times, ergonomics have tried to improve industrial situation by improving the condition of work but workstations and furniture in residential building particularly in old age homes have been neglected to a great extent. Certain research questions like—Are the old age homes build according to the National Building Codes and with the guidelines of ergonomics principles? Are the infrastructure facilities provided appropriate? Are the furniture build according to the Indian furniture standard and by keeping the anthropometric data of elderly people? Are the fittings and fixtures suitable to them? Do they find themselves experiencing any discomfort/pain while using storage spaces and furniture? Are the environmental parameters — temperature humidity, noise and illumination level comfortable to them? Are the constructions of flooring in habitat room and bath non-slippery? and so forth intrigued the investigator. Hence, in a quest to seek answers to some of these queries the present study was envisaged on the topic entitled "ergonomic assessment of old age homes and related health hazards on elderly people" with following objectives:

1. To find out the general profile of elderly people.
2. To study the existing construction and dimensions of old age home in the light of ergonomic principles.
3. To study the infrastructure facilities available in old age home.
4. To find out the satisfaction derived by the elderly people from the old age home.
5. To study related health hazards faced by the elderly people in old age home.
6. To develop a guideline for functional constructive unit for elderly people.

In the present study descriptive-cum-experimental research designs were chosen. Different tools—interview schedule, checklists and experimental tables—were used as instruments for collecting data. Corlett and Bishop CRS used for perceived pain intensity level. A three-point rating satisfaction scale was developed and was judged for content validity and reliability using split-half method.

The purposive random sampling was chosen for selection of area and samples. Descriptive data were collected from randomly selected from four old age homes of Uttarakhand State and out of four old age homes fifteen elderly people were selected randomly thus making total sample of sixty elderly people. Experimental data were collected from twenty per cent of total elderly people. The data were collected in two phases, in first phase descriptive data through interview method, whereas in second phase the experimental data recorded through various instruments.

All the responses received on the data sheet were categorized and analyzed using both descriptive (frequency, percentage, average and standard deviation) and the relational statistics (chi-square, *r*-value, Fisher *t*-test).

Results of the study

Majority of elderly people belonged to 60-80 years age group. The mean age was 72.5 year with 7.18 S.D. Overall number of elderly female was greater than male. Majority of the elderly people were observed to be literate, though the proportion of elderly people by level of education varied, primary, secondary, high school, intermediate, and graduate. A close look at the findings showed that those elderly people in Ist, IIIrd and IVth, *i.e.* paid old age homes were better educated than old age home IInd, while 11.6 per cent illiterate. Most of them were widowers, unmarried females and widows. Majority of elderly people were from general, other backward caste, while a few from SC/ST. None of the elderly people belonged to Muslim community in all the four old age homes. There was preponderance of Christian and Sikh/Punjabi communities in old age homes IIIrd and IVth, respectively. In old age home Ist, IIIrd and IVth the elderly people were independent as these were paid homes, while in IInd majority of elderly people had no income and were dependent on the authority.

Majority of males and females were of the opinion that their existing health status was appropriate according to age and the elderly males were better than females. Majority of them had cataract, weak eyesight and used spectacles. In spite of weak sense of hearing none used hearing aids. Hearing declines more quickly in males than females. Majorities of males were found to have normal taste organs than females. However, the females had better smell organs than males in all the four old age homes. Those who faced walking difficulties took the support of cane and walker. Majority of them had minor health problems, like dental problems, diarrhoea, constipation, bodyache, general weakness, and gaseous distension. Depression was major psychological health problem in both the sexes. Major diseases identified were hypertension, osteoporosis, heart disease, asthma in both the sexes.

Accidents due to slips/falls occurred among elderly people belonging to the age group of 70-80 years. Female elderly people were more prone to accidental falls than male. Many accidental injuries occur in the bathrooms followed by separate toilets and corridors. In most of the cases, the causes of accidents were wet and slippery flooring. The findings of chi-square values showed the dependence of occurrence of accidents on age and sex of elderly people. Majority of female had restriction to daily activities.

The average length, width, and height of habitat rooms in all the four old age home were 4.17, 3.17, and 3.47 metres with standard deviation 1.25, 0.12, and 0.30, respectively. The habitat rooms of all the four old age homes were within the recommended value. The mean length, width, and height of bath and water closet were 1.52, 1.52, and 3.4 metres with standard deviation 0.20, 0.37, and 0.30, respectively. The baths with water closets of all the four old age homes were within the recommended value. The size of separate toilet, bath room and store room was 1.6 × 1.5 metres, 1.5 × 1.0 metres, and 4.1 × 2.3 metres, respectively. The findings of Fisher's *t*-values showed that the dimensions of habitat room were found in accordance with National Building Codes, whereas the separate water closet and store room were different from National Building Codes.

The present study showed the 5th percentile female shoulder height was 117.75 cm. The average maximum height from floor of cupboard was 186 cm. It showed that the majority of elderly people had problems reaching the top shelf of the cupboard of the higher facilities. Though the mean upward reach of the elderly males and females was 217.5 cm, 173.08 cm, respectively for the right hand, most of the elderly females complained of the height of the top shelves.

The mean length and width of door in habitat rooms was 1.64 and 0.79 metres with 0.45 and 0.19 S.D. The average width of the doors of bath with water closets was 0.63 cm which was observed to be far less than the standard recommended width of 80 cm. Hence, it may be caused inconvenience to the elderly people. The mean length, and width of window in habitat rooms was 1.2 × 0.9 metres with 0.21 and 0.08 S.D. The mean length and width of ventilator was 44.75 × 0.92 metres with 3.25 and 0.22 S.D. It was found to be within the recommendation.

The infrastructural facilities in the old age homes were quantified by assigning scores 1 for Yes statement and 0 for No statement, with the total score the percentage was computed. The scores were highest (50 per cent) in old age home IIIrd which showed that the infrastructural facilities were superior followed by old age home Ist, (40 per cent), and in old age home IVth (35 per cent), while lowest (30 per cent) facilities were available in old age home IInd as it was free of cost.

The average total height, seat height from floor, width and

depth of the seat of the chair were 80.4, 45.04, 46.54, and 41.7 cm, respectively. The mean height and width of back rest was 46.6 cm and 45.3 cm, respectively. The materials used for chairs were plastic and wood. The angle of the back rest of wooden chairs was 90°. The mean angle of the back rest was 103.8°.

The average length, width and height from floor of bed were 190, 37, 90.87, and 46.11 cm, respectively with S.D. ranges from 0.13 to 9.75. The computed Fisher's *t*-value showed the significant relationship for height and depth of chair, whereas for width of bed not significant.

The results showed that the 95th percentile of elbow rest height from seat was 23.75 cm and adding ideal seat height to it gives 66.6 cm of table height, while existing table height in all the four old age homes were more than this value. It showed that the existing tables did not match with most of the elderly people.

The mean height of males and females was 154.16 cm and 141.6 cm, with the differences varying from 2.3 to 15.8 S.D. The mean weight of males and females was 68.3 kg and 52.3 kg, respectively with the difference varying from 13.96 to 6.4 S.D. The average body surface area of males and females was 1.16 and 0.96 square metres, respectively. The findings of correlation co-efficient (*r* value) showed significant relations between the height, weight and body surface area of elderly males and females at 5% levels of significance. Anthropometric measurements (standing and sitting posture) of males and females showed highly significant relationship to each other at 1 per cent level of significance.

The average relative humidity level of four old age homes was 67.75 per cent; with 2.62 S.D. falls in humid range but found to be within the recommended value. The average ambient temperature in all the four old age homes was 27.6°C in the month of August 2006. The average noise level in all the old age homes was 41.95 dB; with 3.19 S.D. The average illumination level of four old age homes in habitat rooms was 35.92 Lux with 6.11 S.D. and in bathrooms 21.37 Lux with 3.44 S.D. was found to be far less than the prescribed Indian electrical standard, *i.e.* 50 Lux for habitat room and 100 Lux for bathroom which showed that the old age homes provided insufficient illumination level which may be near fatal to them.

In shoulder external rotation, in using electric switch board required highest average total range of motion value of 62° and in shoulder flexion position, picking articles from floor required highest average total range of motion value of 106.3°. It was observed that the picking articles from floor were most stressful one compared to the other activities. Hence, the articles should be kept within comfortable reach. The normal range of shoulder external rotation = zero to 90 degree and shoulder forward flexion = zero to 180 degree and from the findings it was found in normal ranges.

The angle of the average of the difference in angle, *i.e.* total range of motion value 25° was greater in cervical extension than cervical flexion 8°. The findings depicted that in cervical extension movement (picking article from cupboard upper shelve) was more stressful than in cervical flexion movement (picking article from cupboard lower shelve). The normal range of cervical flexion/ extension = zero to 45 degree and from the findings it was found in normal ranges.

In gross spine measurements the average of the difference in angle, *i.e.* total range of motion value of 5.05° was greater in using storage spaces—lower shelves while using storage spaces—middle shelves (1.87°). It revealed that the using lower shelves were much more stressful than that of using storage spaces—middle shelves.

The findings on pain perception showed that in case of furniture, *i.e.* chair and cot, window, ventilator and while using storage spaces (upper and lower shelves) the elderly people responded up to five points in rating scale of intensity of pain. Mid back pain was found to be a major pain area while using chair. Pain perceived by the elderly people was low when bed used instead of chair. High pain perceived when ventilator was used instead of window, while low when storage spaces (upper shelves) was used instead of lower shelves. Amongst all the possible options explored use of storage spaces (lower shelves) found to be most discomfortable and elevated the perceived pain intensity to maximum in their different body parts. The elderly people faced difficulties because they needed to adopt squatting posture and also used steps to reach to the water closet in old age home IInd. The absence of grab bars in bath with water closets were also complained by the elderly people in all the four old age homes as they faced difficulties in standing up.

For the assessment of risk, the scores earned by elderly people was categorized in three ranges *i.e.* 1-7 scores, 8-15 scores, and 16-23 scores and interpreted as low, moderate and high risk, respectively. Insufficient lightings in bathrooms, stairs, corridors, absence of number of luminaries for light during night hours, lack of generator facilities, lack of bell, absence of grab bars in bathrooms, slippery flooring, improper drainage during rainy season, absence of rubber mats in likely to be wet area were perceived risk by both the sexes. The elderly females perceived higher risk for features like —- unreachable height of ventilators, latches of doors, windows, and storage spaces. Majorities of elderly males were found to be in moderate risk category, while the females were in high risk category. The elderly people belonging to the old age home IInd though were found to be high risks prone yet they affirmed to be satisfied with the available facilities. In old age homes I[st] and IVth the elderly people were satisfied but they perceived higher risk.

In ergonomic assessment, it was found that bath and electric fittings and fixtures, height of windows, ventilators, storage spaces, and the furniture were not suitable as it was not made keeping into consideration the body dimension of elderly people. The flooring of the habitat rooms was not specific, smooth and of grease proof material (skid resistance tiles) particularly in bath with water closets to avoid any accidental slips/falls which was the major drawback. The environmental parameters specially the illuminations level was not friendly to the elderly people and caused accidental injuries which may be near fatal to them. The old age home IIIrd was better as compared to other but still required to be building ergonomically.

Satisfaction scores revealed that the working with ventilator had earned the lowest total score, because the elderly females faced difficulties due to extra height, while daily scheduled activities of the old age home had earned the highest total score. Majority of elderly people were found to be in moderate and high range of satisfaction level, because the old age homes were provided an ambience that will ward off depression and give them a feeling of purpose and belongingness as the authority immerse them in wholesome and worthwhile activities that foster a sense of community among their peers.

It is evident from the chi-square calculated value that in old age homes I[st] and IInd, there was no significance relation between the occurrence of accidents and the age and the sex of elderly people whereas in IIIrd and IVth old age homes the chi-square values showed significant results.

It is evident from the chi-square calculated value that there was no significant relationship between selected anthropometric measurements of elderly people and (a) dimensions of storage spaces and (b) furniture at 5 per cent level of significance.

The height, weight and body surface area of males and females were significantly related to each other at 5 per cent level of significance. Anthropometric measurements (standing and sitting) of males and females showed highly significant relationship to each other at 1 per cent level of significance.

It is evident from the *t*-value that there was no significant relationship between dimensions of habitat rooms and dimensions given in National Building Codes whereas separate water closet and store room showed significant relationship. Hence, it is inferred that the dimensions of habitat rooms and bathrooms with water closets showed equivalence with National Building Codes.

The computed *t*-value showed the significant relationship for height and depth of chair, whereas for width of bed not significant and it is inferred that the furniture provided in old age homes were not made according to the dimensions given for furniture in Indian Standard Codes.

CONCLUSION

On the basis of findings of this investigation the following conclusions were drawn:

1. Elderly people were of the age group of 60-80 years. Female widows dramatically outnumbered male widowers reasons may be that female survival rate is higher than male. The elderly people were literate and from different caste religion.
2. As people age, their strength, endurance, flexibility, vision,

gait, posture, hearing, and balance all tend to deteriorate. Cataracts are another by-product of ageing. Hearing declines more quickly in males than females. The females had better smell organs than males but general health status of males was better than females.

3. All accidents due to slips/falls in the home occurred among elderly people belonging to the age group of 70-80 years. In most of the cases the cause of accidents was slippery flooring. The floor material was found to be substandard, as specific skid resistance tiles are required especially in bathroom but there was ordinary flooring. Moreover, the elderly females were more prone to accidental falls than males. The occurrence of accidents depends on the age and sex of elderly people.
4. The old age homes were not constructed according to the National Building Codes and the guidelines of ergonomic principles.
5. Majority of elderly people had problems reaching the top shelf of the cupboard of the higher facilities. The storage spaces were not made according to the dimensions of human body.
6. The infrastructural facilities in old age home IIIrd were superior to others but still not appropriate and require to be building ergonomically.
7. The furniture, *i.e.* chairs, tables and beds were not constructed according to the anthropometric data and the recommendations given for furniture codes of Indian Standard.
8. There existed significant relationship between the height, weight and body surface area of elderly male and female at 5 per cent levels of significance.
9. There existed highly significant relationship between anthropometric measurement (standing and sitting posture) of elderly male and female at 1 per cent level of significance.
10. The environmental parameters (temperature, humidity, light and sound level) were not comfortable to the elderly people. The illumination level in all the old age homes was found

to be substandard as there was an incandescent fixtures especially in bathroom and it does not provide the required illumination level.

11. Shoulder external rotation/flexion, cervical extension/ flexion and gross spine measurements in the studied group were found in normal recommended ranges.
12. Using storage spaces (lower shelves) found to be most discomfortable and elevated the perceived pain intensity to maximum. Back pain was found to be major pain area of the body while using lower shelves of the cupboard.
13. The elderly people belonging to the old age home IInd though were found to be high risk prone yet they affirmed to be satisfied with the available facilities. In old age home Ist and IVth the elderly people were satisfied but they perceived higher risk. Females were found to be more prone to risk for most of the parameters compared to the males.
14. The old age home IIIrd was superior to other old age homes for the elderly people but require building ergonomically.
15. The elderly people were found to be in moderate and high range of satisfaction level. In old age homes Ist and IVth the elderly people were satisfied but they perceived higher risk.

Hence, it is concluded that all the four old age homes, storage spaces, fittings and fixtures, and furniture were required to be building ergonomically.

RECOMMENDATIONS

The study revealed that all the four old age homes were not ergonomically built, the infrastructure facilities like furniture and furnishing were not appropriate according to their body dimensions and the recommendation given in Indian furniture standard codes. The environmental parameters — temperature/humidity, light and sound level, ill fitting and fixtures in habitat rooms and bath with water closets, and the flooring material had serious shortcomings and can obviously caused adverse health outcome. These risks can be minimized or reduced the severity of others through simple modification in designing the old age homes and carefully selecting the infrastructural facilities, the recommendations are as under:

Design requirements

Elimination of level differences, rounded corners and edges, easily operable facilities like accessible storage spaces (cabinet, cupboard and rack).

Stairs and steps

Stairs and steps both indoors and out are formidable obstacles to many elderly people and are the location of many seriously debilitating accidents. Stairs and steps should be avoided in environments designed for the elderly people or minimized as much as possible and these can be replaced by providing stair lift.

- Handrails should be placed on both sides of the stairs, even within the dwelling unit.
- The rail should extend beyond the steps at the top and the bottom.

Design criteria for architect/builder for storage spaces

Storage-options	*Design Criteria*	*Recommended design values (cm)*
Storage above shoulder height (for light, less frequently used items).	Shoulder height from floor, standing, accessible by 5th percentile female with a 20° (maximum) joint flexion.	117.75 cm.
Shelf height for items requiring visual inspection by the user.	Standing eye height from floor, 5th percentile female.	128.87 cm
Optimal heights of base unit surfaces for elderly males.	Upward reach (50th percentile)	217.5 cm.
Optimal heights of base unit surfaces for elderly females.	Upward reach (50th percentile)	173.08 cm.

Fittings and fixtures

Installation of door level handles, larger and well located electric switches, visual as well as auditory alarm system, sliding doors (particularly for cabinet and cupboard), pull-push handles, and handrails along both sides of building corridors are essential requirements in old age homes.

Bath fixtures

- A raised toilet seat with hand grips on both sides helps. Also, showers should be of non-slip material underfoot and grips at both standing and sitting level.
- The toilet should be positioned next to a sidewall to allow the installation of an assist grab bar.
- Provide a toilet with a seat height of about 18" (the standard is 16").
- Avoid the available high seat toilets that are 20" ; the feet of a short person may not reach the floor.

Lighting fixtures

Good lighting not only assists with reading and safety, it helps prevent depression for those confined indoors.

- A flashlight at bedside and night lights along the route to the bathroom are essential and pay for themselves in accidents prevented.
- If cost is a concern, the improved, inexpensive fluorescent light bulbs may be satisfactory.
- The stairs should be well illuminated by natural or artificial light. A light switch should be located at each end of the stairs. The fixtures should not cause glare, nor should it cause any shadows.
- Wall fixtures are preferred over ceiling fixtures for easier and safer bulb changing. Provide night lights in hallways and bathrooms.

Environmental safety

It can be maximized through sound insulation; improvement in residential temperature control would avert temperature extremes. Maintain a relative humidity of 30-70 per cent at 65-75 degrees Fahrenheit to accommodate a decrease in thermal range tolerance that comes with age.

Furniture and furnishing

Special care should be taken in selection of furniture and furnishing at old age homes.

- The chair redesigning can be done with proper cushioning and back rest with slop of 110^0-115^0 angles with arm and foot rest.
- A redesign of current operating room tables may be required to meet the ergonomic guidelines, table height should be 66.6 cm.
- The bed should not be too low/high, it can be an appropriate height for easy transfer on and off. The metal cots can be replaced with wooden with thick mattresses to avoid back pain.
- The wooden walking cane should be with a crook handle and length adjustable from 74 cm to 97 cm.
- Shoes that are too large, offer poor support, or have high heels can lead to falls, as can loose hosiery and robes or slacks that drag on the floor.
- Clothing that is practical, properly fitting, and conducive to activity is advisable.

Designing of a chair

Actual chair dimensions are determined by measurements anthropometric measurements. For designing the chair ergonomically, the anthropometric data in sitting positions were collected. The chair was designed by Computer Application in Designing at College of Technology, department of Mechanical Engineering, of G.B. Pant University of Agriculture and Technology, Pantnagar (U.S. Nagar), Uttarakhand according to the inputs received from anthropometric data and Indian Standard Codes and this chair is recommended for the elderly people.

Design criteria for furniture manufacturer

Furniture	*Design Criteria*	*Recommended design values (cm)*
Table height	*Seated elbow height or 95th percentile of elbow rest height and adding ideal seat height*	*66.6 cm*
Chair seat height for elderly males	5th percentile popliteal height with an adjustment for shoes.	40.35 + 2.50= 42.85 cm.
Chair seat height for elderly females	5th percentile popliteal height with an adjustment for shoes.	35.75 + 2.50= 38.25 cm
Seat depth	Buttock-popliteal length, 5th Percentile female.	46.07 cm
Seat width	Hip breadth, 95th percentile male.	40.4 cm
Back-rest height above the seat	Shoulder height above seat, 95th percentile male.	65.75
Arm-rest height above the seat	Sitting elbow height above seat, 5th Percentile female.	17.025 cm
Clearance between seat	Buttock to knee length, 95th percentile male + 10 cm for clearance.	56.37 + 10 = 66.37

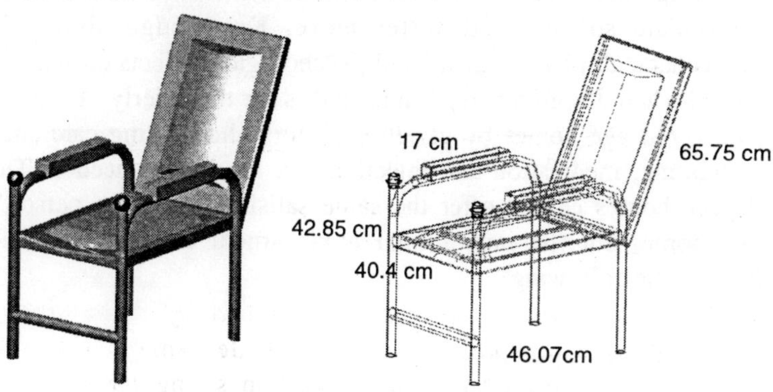

Fig. 5.1 : Ergonomically designed chair

Colour codes

There was lack of colour codes and visual contrast and this can be achieved either through brightness or colour difference to make the key features more perceptible.

- Paint doors and their trim a contrasting colour from the walls. Door thresholds should be flush with the floor, or painted a contrasting colour. Doors left halfway open can be dangerous. One way to solve this problem is with gentle self-closing devices.
- Interior colour schemes with good contrast help elderly people with diminishing eyesight identify doors, entrance ways and changed surface conditions.
- Switch plates, and electrical sockets, that are of contrasting colour from the wall will be easier to see. Use light shades of paint on walls. Flat finishes help eliminate glare.
- Light filtering curtains, blinds, shades and awnings cut down on glare through windows, but allow plenty of indirect sunlight into the house.
- Safety architectural symbols should be marked for electric installation like fuse board, fan, regulator, switch outlets and so forth.

Design consideration should be more for female than male as the female survive and suffer more. Knowledge about the biomechanical, physiological, and psychological effects of aging is needed when designing ergonomic spaces for the elderly. A regular visit to old age homes by geriatric doctors and nursing care such as dressing, medication and injection etc. are highly needed. The old age homes cannot offer the same satisfaction as the person's own home, but the institutional environment can be enhanced through the following:

- An attractive décor.
- Inclusion of the individual's personal possessions.

- Respect for privacy and personal territory.
- Recognition of the individuality of the resident.
- Allowance of maximum control over activities and decision making.
- Environmental modifications to compensate for deficits.
- Social, recreational and educational activities provided to develop their potential to promote a family atmosphere and to enhance mutual support among elderly people.

Colours

Elderly people of old age homes had preferred the under mentioned colours (order of preferences):

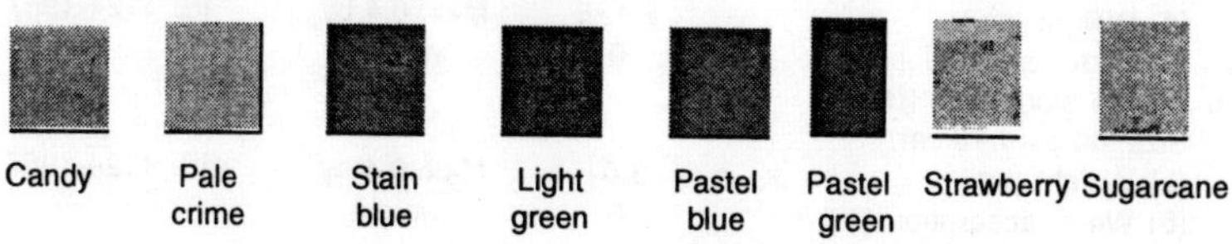

Testing of flooring material

In old age homes the cement concrete and marble stone titles were found in habitat room and bath with water closet. These tiles were taken from the supplier and tested for water absorption level and strength by using compression testing machine in the laboratory of College of Technology G.B.P.U.A.T. Pantnagar. The findings of the test were then compared with the recommended values prescribed in Indian standard codes. It was found suitable for the residential purposes for general populations with regards to the transverse strength.

The floor surface contributes too many trips and falls.

- Mosaic tiles are also popular but should be used on walls instead of floors as they are slippery. Changes in floor levels must be avoided.
- Smooth floor surfaces should be minimized. In potentially wet areas such as kitchens, baths, entries or laundries,

unglazed "Skid Resistance Polycemtiles" which was developed by Central Building Research Institute, Roorkee, (India), and this tile is recommended for the old age homes. The specifications are as under:

Table 5.1 : Testing of flooring material

Characteristics	*Values obtained*	*Recommended values*	*Indian standard codes*
1. Cement concrete tiles			
Size: 250×250×20 (mm)			
(a) Average transverse strength (N/mm^2)	4.75 N/mm^2	0.3 N/mm^2	IS:1237-1980
(b) Water absorption (%)	3.059	10	
2. Marble stone tiles (White)			
Size: 39 cm/ 24.5 cm			
(a) Weight (kg.)	7.125	Max. 0.4 by weight	IS: 1124-1957
(b) Water absorption (%)	0		
3. Marble stone tiles (Black)			
Size: 30 cm / 18 cm			
(a) Weight (kg.)	3.01	Max. 0.4 by weight	IS: 1124-1957
(b) Water absorption (%)	0		

Application/uses:	Suitable for floor and walls of residential, commercial and industrial buildings.
Salient technical features:	Polycem tiles are prepared by combining the cement with the polymers along with some additives for biding the aggregate. There is appreciable improvement especially in physico-mechanical properties like high mechanical and wear strength, low porosity, dimensionally stable, aesthetically appealing and durable.
Environmental aspects:	No adverse effect on the environment.
Level/scale of development:	Laboratory scale.
Status of commercialization:	Ready for commercialization.
Major components/: Raw materials	Polymeric systems compatible with Portland cement, siliceous and calcareous aggregate.
Major plant equipment and machinery:	Mixer, blender, hydraulic press, moulds, polishing equipment.
Techno-economics:	Rs. 10-20 per sq. ft. depending upon texture and colour.
Technology package:	Complete design and construction know-how.

Implications of the study

The demographic ageing of elderly people has implications both on society and economy. It will have an impact at the macro and also at household levels. The magnitude of numbers while encompassing a huge human reserve is also indicative of the scale of endeavours that would be necessary to provide social service and other benefits to the aged persons. The findings of the investigation and the anthropometric data collected in the study brought out a number of implications for concerned users, builders, architects, manufacturers, and equipment designers for a variety of products and facility design applications such as office chairs, easy chairs and tables, bedroom and dinning room furniture, kitchen tops and storage spaces, placement of electric switches, door handles, windows, and so forth, which would help elderly people to work and move about easily and comfortably.

Since large number of builders have entered in the field of construction industry the findings of this study will be of use to them. Large number of resorts are coming up all over India. Thus, the findings of the study may be taken as guidelines by the builders.

Scope for further research

1. Studies can be conducted on the same topic covering
   - Interiors — equipments and products.
   - Effect of various colours on elderly people.
2. A comparative study can be conducted on
   - Rural and urban old age homes of same as well as different States.
3. Studies can be conducted on involving the elderly people in design process:
   - A participatory design model for usability and safety may be developed.

Appendices

Appendix– I

A. INTERVIEW SCHEDULE

General Information

I. Block:
II. District:

1. Name of the Respondent ....
2. Age of the Respondent ....

3. Sex

1. Male
2. Female

4. Education

1. Illiterate
2. Primary school
3. Secondary school
4. High school
5. Intermediate
6. Graduate
7. Post Graduate

5. Marital status

1. Married (with alive spouse)
2. Unmarried
3. Widow
4. Widower
5. Any other

6. Caste

1. General
2. Other backward class
3. Scheduled Tribes
4. Scheduled Caste
5. Any other

7. Religion

1. Hindu
2. Muslim
3. Sikh
4. Christian

8. Financial status

1. No income
2. Retirement pay

9. Place and type of occurrences of accidents

I. Place

1. Toilets (separate)
2. Bathroom with Water Closet

3. Stairs
4. Corridors
5. Room

II. Type

1. Fall
2. Slip
3. Trip

III. Causes of fall

1. Split flooring
2. Slippery flooring
3. Poor light
4. Poor health

IV. Causes of Slip

1. Slippery flooring
2. Wet floor
3. Poor light
4. Poor health

V. Causes of Trip

1. Split floor
2. Uneven surface

10. Restrictions of daily activities due to disease

1. Limited
2. Fairly bit
3. Quite a lot
4. No restrictions

B. HEALTH PROFILE OF THE ELDERLY PEOPLE

1. What is the general status of your health according to you?
   1. Excellent
   2. Appropriate for age
   3. Fair enough
   4. Poor

2. Please tell about the status of your sense organs

   Eyes
   1. Normal
   2. Cataract
   3. Weak
   4. Use spectacles

   Hearing
   1. Normal
   2. Weak
   3. Uses hearing aid

   Taste
   1. Normal
   2. Weak

   Smell
   1. Normal
   2. Weak

   Walking
   1. Normal
   2. Use cane
   3. Use walker

3. Please tell about the minor health problems?
   1. Dental problem
   2. Diarrhoea
   3. Constipation
   4. Bodyache

5. General weakness
6. Depression
7. Gaseous distention

4. Please tell about the major diseases from which you suffer?
    1. Hypertension
    2. Osteoporosis
    3. Heart disease
    4. Asthma
    5. Diabetes mellitus
    6. Cancer

C. CHECKLIST FOR ERGONOMICALLY ASSESSMENT OF OLD AGE HOME

Old Age Home :

Block :

District :

1 = Yes, 0 = No

1. Is the height of habitable room according to National Building Code? ☐
2. Is the height of bathroom/toilet according to National Building Code? ☐
3. Is ceiling sound and free from hazardous defects like damaged, large holes etc.? ☐
4. Are the walls sound and free from hazardous defects like damaged, large holes etc.? ☐
5. Is permanent passageway marked? ☐
6. Are the roof permit effectual drainage of rain water to ensure that the causes of dampness in any part of wall? ☐

7. Are the surfaces of interior finishes on wall generating any type of toxic fume? ☐
8. Are the door and doorways have a clear opening (not less the 80cm) when opened and operable by single efforts? ☐
9. Is the material of doors resistant against moisture? ☐
10. Is there enough number of doors, ventilators in the room to make easy flow of air? ☐
11. Is there any provision for cross ventilation in the room? ☐
12. Is the electrical wiring in a proper condition so as to ensure safety? ☐
13. Are the electric switches arranged in the most favourable locations? (Within the field of vision) ☐
14. Do the safety architectural symbols exist on the electrical installation? ☐

    (Like main fuse board, switch outlets, fan and regulator etc.).
15. Are the stairways safe to walk and non-slippery? ☐
16. Are the bathroom and toilet accessible? ☐
17. Is the minimum fresh air requirement for toilet exhaust (0.60 per metre square of floor area)? ☐
18. Is the floor of the bathroom paved with durable smooth and grease proof material (matt-finished tiles) to avoid any accident? ☐
19. Is the floor of the bathroom sloped towards the drain with a suitable grade? ☐
20. Are floor uniform (one level) in the home or elevated/ depressed? ☐
21. Is the corner between the walls rounded? ☐
22. Is the grip part of the handle of door comfortable? ☐
23. Is the height of windows and ventilators suitably design for you? ☐
24. Are the storage spaces (cupboard, rack, etc.) suitable to you? ☐

25. Is the height of the table comfortable to you? ☐
26. Are you using a lower height chair/sofa which causes difficulty in standing up? ☐
27. Is the bed at an appropriate height for easy transfer on and off? ☐
28. Are the standard fixtures and fitting (Taps, Cupboards) easy to access? ☐

D. CHECKLIST FOR INFRASTRUCTURE FACILITIES AVAILABLE IN OLD AGE HOME

Yes = 1 No = 0

I. Furniture and furnishing

1. Is there sufficient number of furniture? ☐
2. Is there any special kind of easy chair for relaxing? ☐
3. Are the mattress and cushion comfortable in respect to texture and colour? ☐
4. Is there dinning table facility? ☐
5. Are there rubber mates placed in areas likely to get wet? ☐

II. Fitting and fixtures

6. Is there special kind of handholds in sleeping, bathroom, toilet spaces? ☐
7. Is the lighting adequate around stairwells, bathroom and working areas? ☐
8. Are there larger switches which glow-in-light and visible from the colour of the wall ☐
9. Is there sufficient number of wash-basins? ☐
10. Is the toilet facility adequate for the number of inmates? ☐
11. Are there any special toilet seats with arm rest or side bar etc.? ☐

III. Medical facilities

12. Is the personal emergency response system provided for immediate assistance in the event of an emergency (Bell)? ☐
13. Is there regular health check-up (Temperature/Blood pressure) facilities? ☐
14. Is there first aid facility for urgency? ☐
15. Are there any facilities for free medicine? ☐
16. Is there ambulance facility for an emergency? ☐

Any other facilities

17. Are closed containers provided for waste material disposal? ☐
18. Are there sufficient recreational facilities — T.V., radio, library, and temple? ☐
19. Is there any facility at the time of light cut off? (Generator) ☐
20. Are there phone facilities in room? ☐

E. ASSESSMENT OF PERCEIVED PAIN

1. Do you feel any type of pain/numbness or any other type of symptom while using the following in habitat room?

Yes/No ☐

(If yes, then where?)

Body parts	*Chair*	*Bed*	*Window*	*Ventilator*	*Shelves of cupboard*
• Neck					
• Shoulder					
• Elbow/Forearm					
• Hand/wrist					
• Back					
• Knee					
• Ankle					

Give Reasons

1 ____________

2 ____________

3 ____________

3. Do you feel any type of pain/numbness or any other type of symptom while using bathroom with water closet?

Yes/No ☐

(If yes then where?)

Body parts	*Ventilator*	*Water closet*
• Neck		
• Shoulder		
• Elbow/Forearm		
• Hand/wrist		
• Back		
• Knee		
• Ankle		

Give Reasons

1 ____________

2 ____________

3 ____________

F. POSTURAL DISCOMFORT—COMBINED RATING SCALE

In this picture you can see the approximate position of the parts of the body referred to in the questionnaire. You should decide for yourself in which part you have or have had your trouble.

0 Nothing at all
0.5 Extremely weak
1. Very weak
2. Weak (light)
3. Moderate
4. Somewhat strong
5. Strong (heavy)
6. In between strong and very strong
7. Very strong
8&9. In between strong and extremely strong
10. Extremely strong

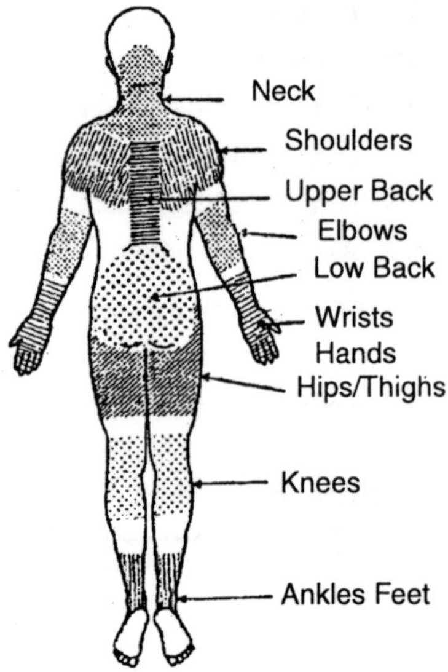

G. CHECKLIST FOR ASSESSMENT OF RISK IN OLD AGE HOME

The following statements are related to existing conditions; indicate whether you have ever faced any problems by stating Yes/No

Yes = 1, No = 0

1. Insufficient lighting in bath, stairs, corridors ☐
2. Absence of night light ☐
3. Absence of generator facility ☐
4. Absence of bell for emergency ☐
5. Unreachable height of door, window latches ☐
6. Unreachable height of ventilator ☐
7. Inaccessibility of bath with water closet ☐
8. Squatting toilet (lower seat) ☐

9. Absence of grab bars in bathroom ☐
10. Slippery and uneven flooring ☐
11. Inaccessibility of dinning spaces ☐
12. Height of storage spaces in room ☐
13. Height of storage spaces in bathroom ☐
14. Absence of handrails at the top and bottom of stairs ☐
15. Obstructed pathways ☐
16. Raised structure/equipment ☐
17. Absence of rubber mats in wet area ☐
18. Bolts/nuts/sharp edges of furniture ☐
19. Improper drainage ☐
20. Improper clothing/footwear ☐
21. Steps too high ☐

H. SATISFACTION SCALE

Name of the respondent:

Old age home:

Statements	*Highly Satisfied*	*Satisfied*	*Not satisfied*
Score	3	2	1

Are you satisfied with the following statements?

1. Size of room ☐
2. Size of door ☐
3. Size of window ☐
4. Size of ventilator ☐
5. Height of cupboard ☐
6. Height of shelve ☐
7. Working with door ☐

8. Working with window ☐
9. Working with ventilator ☐
10. Direction/placement of door ☐
11. Direction/placement of window ☐
12. Direction/placement of ventilator ☐
13. Handles of doors and window ☐
14. Condition of walls, ceiling ☐
15. Condition of floor ☐
16. Number of furniture in the room ☐
17. Size and shape of chair ☐
18. Size and shape of table ☐
19. Size and shape of bed ☐
20. Mattress and cushions ☐
21. Size of bathroom and W.C. ☐
22. Internal temperature ☐
23. Sanitary conditions (waterlogging) ☐
24. Noise level ☐
25. Daily activities ☐
26. Number and type of luminaries ☐
27. Natural lighting ☐
28. Placement of electric switches ☐
29. Recreational facilities ☐
30. Medical facilities ☐

Aappendix – II

RECORDING SHEET - 1

Old Age Home :

Block :

District :

Dimensions of the old age home

Descriptions	*Length*	*Width*	*Height*
Old age Home I			
Habitat room			
• Dinning room			
• Bath with W.C.			
• Any other places			
Old age Home II			
• Habitat room			
• Dinning room			
• Bath with W.C.			
• Any other places			
Old age Home III			
• Habitat room			
• Dinning room			
• Bath with W.C.			
• Any other places			
Old age Home IV			
• Habitat room			
• Dinning room			
• Bath with W.C.			
• Any other places			

RECORDING SHEET - 2

Dimensions of doors, windows, ventilators and storage spaces in the old age homes

Descriptions	*Number*	*Height*	*Width*	*Depth used*	*Material*
1. Habitat room					
Door					
Window					
Ventilator					
Cupboard					
Shelve					
Rack					
2. Dinning room					
Door					
Window					
Ventilator					
Cupboard					
Shelve					
Rack					
3. Bath with W.C.					
Door					
Window					
Ventilator					
Cupboard					
Shelve					
Rack					

RECORDING SHEET - 3

Dimensions of furniture

Descriptions	*In old age home I*	*In old age home II*	*In old age home III*	*In old age home IV*
Dimensions of Chair				
Total height				
Seat height				
Seat depth				
Seat width				
Back rest height				
Back rest width				
Angle between seat and back rest				

Material used

Dimensions of Table
Table length
Table depth
Height of knee zone

Dimensions of bed
Bed length
Bed width
Height from floor
Any other

RECORDING SHEET - 4

Recording of anthropometric measurement of the respondents

Name of the respondent:
Age :
Height :
Weight :
Body surface area :

Anthropometric measurements (sitting positions)

1. Body height from ground ☐
2. Body height above seat ☐
3. Eye height from ground ☐
4. Eye height above seat ☐
5. Shoulder height from ground ☐
6. Shoulder height above seat ☐
7. Elbow height from ground ☐
8. Elbow height above seat ☐
9. Knee height above ground ☐
10. Popliteal height above ground ☐
11 Buttock popliteal lengths ☐

12. Buttock knee lengths ☐
13. Thigh heights ☐
14. Hip breadth ☐

Anthropometric measurements (standing positions)

1. Body weight ☐
2. Eye height ☐
3. Shoulder height ☐
4. Elbow height ☐
5. Knee height ☐
6. Popliteal height ☐
7. Upward reach ☐
8. Sideways reach ☐
9. Forward reach ☐
10. Shoulder-elbow height ☐
11. Forearm-elbow height ☐
12. Hand length ☐
13. Elbow to elbow breadth ☐
14. Total arm length ☐

RECORDING SHEET - 5

Recording of ambient temperature, relative humidity, light and sound levels

Rooms	*Mean*	*S.D*
1. Ambient temperature (°C)		
• Habitat room		
2. Relative humidity (%RH)		
• Habitat room		
3. Light level (Lux)		
• Habitat room		
• Bath with Water Closet		
4. Sound level (Decibel)		
• Habitat room		

RECORDING SHEET - 6

(a) Upper extremities shoulder external rotation/flexion

Type of activity	*Primary Sensor (°) Mean Repetitions*			*Total Range of Motion*
	I^st	*IInd*	*IIIrd*	*Mean (°)*
External Rotation				
1. Opening/closing door latches				
2. Opening/closing window				
3. Using switch board				
4. Using platform for keeping articles				
5. Using T.V./music system				
Shoulder Flexion				
6. Using platform for keeping articles				
7. Using lower shelve of cupboard				
8. Picking articles from floor				

(b) Upper extremities cervical flexion/extension

Type of activity	*Primary Sensor ROM Value(°)*		*Secondary Sensor ROM Value (°)*		*The difference in angle Total ROM (°)*	
	Repe-tition	*Mean*	*Repe-tition*	*Mean*	*Repe-tition*	*Mean*
Cervical Flexion Picking articles from cupboard (lower shelve)	I^st IInd IIIrd					
Mean						
Cervical Extension Picking articles from cupboard (upper shelve)	I^st IInd IIIrd					
Mean	☐					

(c) Gross spinal flexion range of motion measurements while using storage spaces

Type of activity	*Repetition*	*Primary sensor (°)*	*Secondary sensor (°)*	*Total range of motion (°)*
Using storage spaces in the room (lower shelve)	I[st] IInd IIIrd			
Mean				
Using storage spaces in the room (middle shelve)	I[st] IInd IIIrd			
Mean				

Bibliography

Agarwal, B.L. (2002). *Basic Statistics*, 3rd edition, New Age International (P) Limited, Publishers, New Delhi, pp. 243-248.

Allan, J., Khong, C.W., Gilhaum, B., Hall, S., Kerwood, J., Macdonald, A., McNally, N., Nelson, D., Page, S., Stewart, S., Stovall, D., (1996). *The Challenge of Age*. Glasgow School of Art: The Foulis Press.

Bagga, Amrita. (1997). *"A study of women in old age homes of Pune"*, In : Chakravarty Indrani (ed.), *Life in twilight years*. Calcutta: Quality Book Company, 1997, pp.171-175.

Bailey, I. (1995). "Visual Environments of Older Adults". In D. Driver (Ed.). *Proceedings of Blueprint for Ageing*. Oakland, CA.

Banerjee, S. and Sen, R. N. (1955). "Nomogram for calculating Surface Area". *Journal of Applied Physiology* 7, p. 583.

Barbaccia, J. C. (1995). "Changes in Physical Function and the Ageing Process: Implications for Facility Design". In D. Driver (Ed.). *Proceedings of Blueprint for Ageing*. CA, Oakland.

Barton, N. J. (1992). "Occupational causes of disorders in the upper limb". *British Medical Journal 304*: 309-311.

Benktzon, M. (1993). "Designing for Our Future Shelves: The Swedish Experience". *Applied Ergonomics* 24 *1:* 19-27.

Bhandari, A. (1992). "CIB W84 Report: Access Design and Legislation", CIB Expert Seminar on Building Non-Handicapping Environments, Harare: Institute on Independent Living.

Boess, S., Lebbon, C., and Maggs, D. (1999). "A Case of Cooperation Conference between Designers and Older Consumers". *Fourth Global Conference of the International Federation on Ageing Book of Abstracts*. Montreal: 39.

Borger, L.L., Whitney, S.L., Redfern, M.S. and Furman J.M. (1999). "The Influence of Dynamic Visual Environments on Postural sways in the elderly". *Journal of Vestibular Research* 9:197-205.

Borkan, G.A., Hults, D.D. and Glynn. R.J. (1983). "Role of longitudinal change and secular trend in age differences in male body dimensions". *Human Biology 55*: 629-641.

Boyce, P.R., Akashi, Y., Hunter, C.M. and Bullough, J.D. (2003). "The Impact of Spectral Power Distribution on the Performance of an Achromatic Visual Task", *Lighting Research and Technology*, *(35)*, pp. 141-161.

Brink, S. (1996). "Housing Options for Elderly People: Generic Solutions that Include or Special Solutions that Exclude?" In E. Komut (Ed.). *Housing Questions of the 'Others'*. Ankara: Chamber of Architects of Turkey, pp. 347-356.

Cavanagh, S. (1996). "The Space we need: Principles of Housing Design for Older Women, Women with Children, and Parents with Disabilities". In E. Komut (Ed.). *Housing Questions of the 'Others'*. Ankara: Chamber of Architects of Turkey, pp. 67-76.

Chadha, N. K. and Skogstad, A. (1998). "Psychosocial determinants of Institutionalized elderly. An empirical study". *Indian Journal of Gerontology 12 (3, 4)*: 27-39.

Coleman, R. (1997). "Breaking the Age Barrier. Speech at the Royal Society of Arts, June '97 as part of the Design Council" - Design in Education Week, London: Design Age, Royal College of Art, March 12th http://valley.interact.nl/DAN/ NEWSLETTER/ NEWS97/ home.html,6, 6, 1998.

Collins, K.J. (1993). "Cold and heat-related illnesses in the indoor environment". In R. Burridge & D. Ormandy (Eds.), *Unhealthy Housing: Research, Remedies, and Reform,* pp. 117-137. London: E. and F.N. Spon London.

Coni, N., Davison, W. and Webster, S. (1992). *Ageing: The Facts.* Oxford, N. York: Oxford Univ. Press. *CybErg'96.*http:// www.curtin.edu. au/conference/cyberg/center/paper/demirbilek/ paper.htlm, Perth 1-30 September.

Damon, A. and Stoudt, H.W. (1963). "The functional anthropometry of old men". *Human Factors 5*: 485-491.

Dandekar, Kumudini. (1996). *Elderly in India*. Sage Publications, 1996. New Delhi-7, p. 229.

Demirbilek, O. (1996b). "Ageing and the Activities of Daily Living". *Proceedings of the 1st International Virtual Conference on the Internet on Ergonomic Design, Research, Education*. Newcastle: Athenaeum Press, pp. 193-206.

Du Bois, D. and Du Bois, Errors F. (1916). *Arch. Int. Med.*17, p. 863.

Enquselassie, F., Dobson, A.J., Alexander, H.M. and Steele, P.L. (1993). "Seasons, temperature and coronary disease". *International Journal of Epidemiology.* 22 : 632-636.

Esmen, N. A. (1985). "The status of indoor air pollution". *Environmental Health Perspectives 62:* 259-65.

EUFRP, (1997). PRESENCE Project, European Union Funded Research Programme, http://www.presenceweb.org, 10, 3, 1998.

Evans, W.A., Courtney, A.J. and Fok, K.F. (1988). "The design of school furniture for Hong Kong School Children", *Applied Ergonomics. 19 (2)* : 122-134.

Galer, I.A.R. (1995). *Applied Ergonomics Handbook*, Butterworth, London, pp. 103-119.

Genaidy, A.M. and Karwowski, W. (1993). "The effect of neutral posture deviation on perceived joint discomfort ratings in sitting and standing postures". *Ergonomics* 36: 785-792.

Gill, T. M., Williams C.S., Tinetti, ME. (2000). "Environmental hazards and the risk of non-syncope fall in the homes of community-living older persons". *Medical Care 38 (12)*: 1174-1183.

Gitlin, L. (1998). "Testing Home Modification Interventions: Issues of Theory, Measurement, Design and Implementation". *Annual Review of Gerontology and Geriatrics 18*:190-237.

Grandjean, E. (1973). *Ergonomics of the Home*. Taylor and Francis, London, pp. 287-307.

Grandjean, E. (1981). *Fitting the task to the man: A textbook of occupational ergonomics*. London, UK: Taylor and Francis.

Grandjean, E. (1988). *Fitting the task to the man: A textbook of occupational ergonomics* (4th Ed). London, UK: Taylor and Francis. pp. 35-36.

Grozdanovic, M. (2002). "Human activity and musculoskeletal injuries and disorders". *Medicine and Biology* 9 *(2)*: 150-156.

Harold, G.L. (2000). *ASHRAE Handbook—HVAC Systems and Equipment.* Atlanta: American Society of Heating, Refrigerating and Air-conditioning Engineers, Inc., pp. 201-256.

Harving, H., Korsgaard, J. and Dahl, R. (1994). "Clinical efficacy of reduction in house-dust mite exposure in specially designed, mechanically ventilated healthy homes". *Allergy 49 (10)*: 866-870.

Haslegrave, C.M. (1994). "What do we mean by a working posture"? *Ergonomics 37: 4:* 781-799.

Helms, R.N. (1980). *Illumination Engineering for Efficient Luminous Environments,* Engleword Cliffs: Prentice- Hall, Inc., p. 5.

Hopton, J. and Hunt, S. (1996). "The Health Effects of Improvements to Housing: A Longitudinal Study". *Housing Studies 11 (2)*: 271-286.

Howe, D.A. (1992). "Creating Vital Communities: Planning for Our Ageing Society". *Planning Commissioners Journal,* December: 7.

IS: 5974-1986 (Reaffirmed 1996). *Indian standards specification of divans and easy chairs (First Revision),* Indian Standard Institution, New Delhi, p. 6.

IS: 7259 (Part 1)–1988. (Reaffirmed 1996) *Indian standards specification for Wooden Beds* Indian Standard Institution, New Delhi, p. 2.

Kellog, F., Bertolone, S. and Bertolone, F. (1986). "Bringing Interiors to Light: The Principles and Practices of Lighting Design", In Diane Casella Hines (Ed.). New York: Whitney Library of Design, Watson-Gupill Publications: *14*: 176-180.

Kinsey, J. and Lane, S. (1983). "Race, housing attributes, and satisfaction with housing". *Housing and Society 10 (3)*: 98-116.

Kose, S. (1993). "Changing Lifestyles, Changing Housing Form:

Japanese Housing in Transition". In M. Bulos and N. Teymur (Edn), *Housing: Design Research, Education.* Newcastle: Athenaeum Press, pp. 193-206.

Kose, S. (1996). "Ageing in Place: Importance of Universal Design Concept in Housing Policy". E-mail on *ISA RC43 Conference*, skose@keken.go.jp, August 11th, 1996.

Kothiyal, K. Tettey, S. (2001). "Anthropometry for design for the elderly". *International Journal of Occupational Safety and Ergonomics 7 (1)*: 15-34.

Lowery, S. (1989). "Indoor air quality". *British Medical Journal,* Vol. pp. 1338-40.

Mant, D. (1993). "The parameters of health and housing research". In R. Burridge & D. Ormandy (Eds.), *Unhealthy Housing: Research, Remedies, and Reform* London: E. & F.N. Spon. pp. 1-21.

Markus, T.A. (1993). "Cold, condensation and housing poverty". In R. Burridge & D. Ormandy (Eds.), *Unhealthy Housing: Research, Remedies, and Reform.* London: E. & F.N. Spon. pp. 141-165.

McCray, J.W. and Day, S. (1977). "Housing values, aspirations, and satisfactions as indicators of housing needs". *Home Economics Research Journal* 5 *(4)*: 244-254.

Mishra, S. (1993). "Scenario of segregated seniors. A Case study of inmates of old age homes". *Indian Journal of Gerontology 7(3,4)*: 104-110.

Mood, E.W. (1993). "Fundamentals of healthful housing: Their application in the 21st century". In R. Burridge & D. Ormandy (Eds.*), Unhealthy Housing: Research, Remedies, and Reform,* London: E. & F.N. Spon. pp. 303-334.

Morini, A. and Pomposini, R. (1996). "New Designs for Aged People Housing". In O. Ural, D. Alt_nbilek, and T. Birgtnül (Eds.). *XXIVth IAHS World Housing Congress,* May 27-31, Ankara: Middle East Technical University, pp. 791-800.

Nalini, B. (1997). "Institutional care for the aged. The issue and implications". In : Chakravarty, Indrani (ed.), 1977. *Life in twilight years*, New Delhi: Quality Book Company, 1997, pp. 63-77.

Nicol, J.F. (1993). *Thermal Comfort: A Handbook for Field Studies towardsan Adaptive Model.* London: University of East London.

Noble, B.J., Borg, G.A.V., Jacobs, L.,Ceci, R.and Kaiser, P. (1983). "A category ratio perceived exertion scale: relationship to blood", *Medicine and Science in Sports and Exercise, 15*: 523-526.

Pheasant, S. (1988). *Body space: Anthropometry, Ergonomics and Design,* (2nd Edition), London: Taylor and Francis.

Pheasant, S.T. (1986). *Body Space: Anthropometry, Ergonomics and Design* (2nd Edition), London, UK: Taylor and Francis.

Pheasant, S.T. (1996). *Body Space: Anthropometry, Ergonomics and the Design of Work* (2nd Edition), London: Taylor and Francis.

Pitt, B. (1982). *Psycho geriatrics, an introduction to the Psychiatry of Old Age*. Edinburgh: Churchill Livingstone.126.

Rajan, S., Mishra, U.S. and Sharma, P.S. (1999). *India's elderly Burden or Challenge?* New Delhi: Sage Publications, 1999.

Ranzijn, R. (2002). "The Potential of Older Adults to Enhance Community Quality of Life: Links between Positive Psychology and Productive Ageing", *Ageing International* 27 *(2)*: 30-35.

Raschko, B. B. (1991). *Housing Interiors for the Disabled and Elderly*. New York : Van Nostrand Reinhold, p. 48.

Richardson, S. and Poulson, D. (1996). *Usability-Disability and Elderly*. 3rd Annual Concertation Meeting of the Telematics Applications Programme, Applications for the European Information Society. Brussels: Palais des Congrés, Dec. 2-3.

Roebuck, J., Kroemer, K. and Thompson, W. (1975). *Engineering anthropometric methods*, New York, N.Y., USA: Wiley.

Rudge, A. (1997). "Access Prohibited: Information for Designers of Public Access Terminals". http://www.eyecue.co.uk/pats/ contents.html,8, 3,

Salmon, G. (ed.) (1993). *Caring Environments for Frail Elderly People*. Essex: Longman Scientific and Technical, New York: J. Wiley and Sons, pp. 8-11.

Samet, J.M., Marbury, M.C. and Spengler, J.D. (1988). "Health effects and sources of indoor air pollution". Part II. *American Review of Respiratory Diseases, 137*: 221-242.

Schneider,B. and Pichora-Fuller. (2000). Implications of Sensory Deficits for Cognitive Aging in *The Handbook of Aging and Cognition*, 2nd ed., F.I.M. Craik and T. Salthouse, Eds., Lawrence Eri-baum Associates Mahwah, NJ, pp.155-219.

Scott Woodcock (2005). *Australia's guide to environmentally sustainable homes. Third edition, A joint initiative of the Australian Government and the design and construction industries. Your Home Technical Manual - 6_3 Health and Safety.htm*

Shah, B. (1987). *Socio-economic scale for rural-urban group*, Psychological Research Cell, Agra.

Sheehan, N.W. (1992). *Successful Administration of Senior Housing: Working with Elderly Residents.* Newbury Park, CA; London: Sage Publications.

Snedecor, G. W. and Cochran, W.G. (1968). *Statistical Methods*. 6th edition, Lown State University Press: Ames. Lowa, p. 558.

SP7: (1983). *Handbook : National Building Codes of India. 1993.* Published by Bureau of Indian Standards, New Delhi. Group I, pp. 17-31.

Steenbekkers, L.P.A. and Beijsterveldt, C.E.M. (1998). *Design-relevant characteristics of ageing users.* Backgrounds and guidelines for product innovations. Delft. The Netherlands: Delft University Press.

Stoudt, H.W. (1981). "The anthropometry of the elderly". *Human Factors 23:* 29-37.

Sturges, I., Rudd, A.G. and Shilling, J. (1994). "Unrecognized Visual Problems amongst Residents of Part III Homes". *Age and Ageing. 23*: 54-56.

Varghese, M. A., Atreya, N. and Bhatnagar, A. (1989). *An experimental to determine convenient height for kitchen activities.* U.G.C. Report 9. Bombay, S.N.D.T, Women's University. p. 9.

Vatuk, Sylvia, (1990). *To be a burden on others. Dependency anxiety among the elderly in India.* In: Lyuch Owen (ed.), 1990. Divine passions. The social construction of emotion in India, Berkeley: University of California Press, 1990, pp. 64-88.

Walker, B. Jr. (1990). "Housing and Health". *Bulletin of the New York Academy of Medicine, 66 (5)*: 382-90.

Wolff, C. (1997). A Haven, Not a Hospital: Assisted Living Design Involves Practical and Emotional Issues. *Lodging Hospitality.* November 1st, *53 (3)*: 64.

Woodson, W. (1981). *Human factors design handbook.* McGraw-Hill, New York.

World Health Organization. (1987). *Housing: The implications for health.* Geneva: World Health Organization.

Woudhuysen, J. (1993). "A Call for Tran generational Design". *Applied Ergonomics 24 (1)*: 44-46.

Wright, F. (1994). *Accident Prevention and Risk-taking by Elderly People: The Need for Advice,* United Kingdom: Department of Trade and Industry.

Index